INTO[illegible] OF ALUMINUM ON BIOCHEMICAL AND HISTOPATHOLOGICAL

By. GANAPATHY, M.Sc.

ACKNOWLEDGEMENT

I wish to express my deep sense of gratitude to my supervisor Dr.N.Jothi Narendiran, M.Sc.,M.Phil.,B.Ed.,Ph.D., Assistant Professor in Zoology, PG and Research Department in Zoology, Government Arts College for Men, Nandanam, Chennai-600035 for his guidance to chose the research plan and towards the successful completion of this work.

It is with great pleasure, I take this opportunity to express my thanks to Dr. A. Yousuf Khan, M.Sc.,M.Phil.,Ph.D., Associate Professor and Head, PG and Research Department of Zoology, Government Arts College for Men, Nandanam, Chennai-600035 for the support rendered.

I wish to thank our Principal Dr. R.Jayachandran, M.A., M.Phil., Ph.D., Government Arts College for Men, Nandanam, Chennai-600035, for permitting me to carry out my research work.

I owe my deep sense of gratitude to Dr. (Mrs.) R. Usha, Professor and Head (Retd.) Dept. of Zoology, Presidency College (A), Chennai. For her guidance and intellectual criticisms throughout my tenure.

I would also place my heartfelt thanks to Dr. K. Elumalai, Assistant Professor, Doctoral Committee Member, PG and Research Department of Zoology, Government Arts College for Men, Nandanam, Chennai-600035, for helping me to complete my research.

I am extremely thankful to Dr. M. Majeetha Parveen, Associate Professor, Prof. J. Richardson, Associate Professor, Dr. P. Martin, Assistant Professor, Dr. T. Ramachndran, Assistant Professor, PG and Research Department of Zoology, Government Arts College for Men, Nandanam, Chennai-600035, for their valuable suggestions and timely help rendered

.

It is my great pleasure to acknowledge my deep sense of gratitude and heartfelt thanks to, Dr. R Krishnamurthy, Dr. G. Dalmin., Assistant Professors, Department of Zoology, Govt. Arts College (Autonomous), Nandanam, Chennai – 35, for their valuable guidance and constructive criticism during the tenure of the present investigation.

It is my privilege to thank Dr. (Mrs.) R. Jothi, , Dr.B. Venkadesh, Dr. Winkins Santosh, Dr. M. Elumalai Dr. K. Maheshkumar, Prof. (Mrs.) Uma Maheshwari,, Dr. A. Manimaran, Assistant Professors, Department of Zoology, Govt. Arts College (Autonomous), Nandanam,Chennai-35, for their support extended to me for carrying out the statistical analysis of my data.

I would like to extend my sincere thanks to. Mr. N. Sivakumar and Mrs. S. Manjula, Chellampattidai village, and my family Members for their timely and valuable help.

I would like to place my thanks to my beloved Parents for the support rendered by them for building up my career and the motivation given by my siblings Mr. A. Nagappan and Mrs. N. Muththammal, Chellampattidai.

I Express my thanks to my beloved Uncles Mr. P. Selvaraj Vasantha, (Rited), Fishery Department, Kancheepuram, P. Mahendran Deisi, Tailor shop, Mosur and P. Saravanan Sumathi Tailor shop, Chellampattidai for his showers of blessings which enabled me to complete this research work.

Words would not suffice to thank my Sister Mrs. N. Vimala Ranjan, R. Vimal BA., Kalambakkam and for their understanding and immense patience shown towards me throughout my research work, they have been a source of inspiration throughout the course of this study.

List of Tables

Table 13. Leucocytes count into the blood of *Cyprinus carpio var. communis,* exposed in the direction of sub-lethal concentration of Aluminium sulphate used for varying periods.

Table 14. MCHC values of *Cyprinus carpio var. communis,* exposed in the direction of sub-lethal concentration of Aluminium sulphate toxicity used for varying periods.

Table 15. MCH value of *Cyprinus carpio var. communis,* exposed in the direction of sub-lethal concentration of Aluminium sulphate toxicity used for varying periods.
Table 16. MCV values of *Cyprinus carpio var. communis,* exposed in the direction of sub-lethal concentration of Aluminium sulphate toxicity used for varying periods.

Table 17. Protein in addition to glucose levels into the plasma of *Cyprinus carpio var. communis,* exposed in the direction of acute Aluminium sulphate toxicity.

Table 18. Protein level into the plasma of *Cyprinus carpio var. communis,* exposed in the direction of sub-lethal Aluminium sulphate toxicity used for varying periods.

Table 19. Glucose level into the plasma of *Cyprinus carpio var. communis,* exposed in the direction of sub-lethal Aluminium sulphate toxicity used for varying periods.

Table 20. Na^+, K^+, ATPase activity into the gills in addition to plasma also plasma phosphatases movement of *Cyprinus carpio var. communis,* expose in the direction of acute Aluminium sulphate toxicity.

Table 21. Na^+, K^+, ATPase activity into the gills of *Cyprinus carpio var. communis,* exposed in the direction of sub-lethal concentration of Aluminium sulphate toxicity used for varying periods.

Table 22. Na^+, K^+, ATPase activity into the plasma of *Cyprinus carpio var. communis,* exposed in the direction of sub-lethal concentration of Aluminium sulphate toxicity used for varying periods.

Table 23. Acid phosphatase activity into the plasma of *Cyprinus carpio var. communis,* exposed in the direction of sub-lethal concentration of Aluminium sulphate used for varying periods.

Table 24. Alkaline phosphatase activity into the plasma of *Cyprinus carpio var. communis,* exposed in the direction of sub-lethal concentration of Aluminium sulphate toxicity used for varying periods.

List of Figures

Fig.13. MCHC values of *Cyprinus carpio var. communis,* exposed to sub-lethal concentration of Aluminium sulphate used for varying periods.

Fig.14. MCH values of *Cyprinus carpio var. communis,* exposed into the direction of sub-lethal concentration of Aluminium sulphate used for varying periods.

Fig.15. MCv values of *Cyprinus carpio var. communis,* exposed in the direction of sub-lethal concentration of Aluminium sulphate used for varying periods.

Fig.16. Protein in addition to glucose levels into the plasma of *Cyprinus carpio var. communis,* exposed in the direction of acute Aluminium sulphate toxicity.

Fig.17. Protein level into the plasma of *Cyprinus carpio var. communis,* exposed in the direction of sub-lethal concentration of Aluminium sulphate used for varying periods.

Fig.18. Glucose level into the plasma of *Cyprinus carpio var. communis,* exposed in the direction of sub-lethal concentration of Aluminium sulphate used for varying periods.

Fig.19. Na^+, K^+, ATPase activity into gills and plasma and plasma phosphatase activity of *Cyprinus carpio var. communis,* exposed to 24 hours LC_{50} concentration of aluminium sulphate toxicity.

Fig.20. Na^+, K^+, ATPase activity into the gills of Cyprinus carpio *var.* communis, exposed to sub-lethal concentration of Aluminium sulphate toxicity used for varying periods.

Fig.21. Na^+, K^+, ATPase activity into the plasma of *Cyprinus carpio var. communis,* exposed in the direction of sub-lethal concentration of Aluminium sulphate toxicity used for varying periods.

Fig.22. Acid phosphatase activity into the plasma of *Cyprinus carpio var. communis,* exposed to sub-lethal concentration of Aluminium sulphate toxicity used for varying periods.

Fig.23. Alkaline phosphatase activity into the plasma of Cyprinus *carpio var. communis,* exposed in the direction of sub-lethal concentration of Aluminium sulphate toxicity used for varying periods.

List of Plates

INTRODUCTION

Qualitative aspect of toxicology are significant as they are basic to the security estimate procedure into which, individual first determine the toxicological profile of the material and next establish how the chemical can be there employ carefully to stop injury (Plaa, 1982). According in the direction of Durham (1974), toxicity is the skill of a substance particle or compound in the direction of create injury just the once it reach susceptible place which is determined by means of the dose. Cairns (1984) report to starting a regulatory tip of view, toxicity tests are use for three main purpose; they are (i). Showing of chemicals also harvest, (ii). Establish limit and (iii). Monitor; the writer additional stated to bioassay analysis can be use to establish the highest satisfactory application of a pollutant into a particular atmosphere lacking deliberate appliance of the chemical cause some bad natural consequence.

Ramade, (1987), inside his analysis acknowledged that many specialist contain alert their concentration going on the difficulty of bio-assay in the monitor of worldly with aquatic ecosystem. The plan of use bio-assay in monitor of ecological pollution is to create a association involving toxicity in addition to absorption of a pollutant organism study inside the biotope; the contaminated effects be able to exist separated keen on two category *viz.,* effect to arise extremely fast behind a short contact to a substance agent (Acute) with persons to come into view single following recurring contact to the substance (Chronic), (Durham. 1974; in addition to Nagel, 1993).

Hagen (1959) pointed out the objectives of acute toxicity studies

- To evaluate toxicity into non-target species,
- To predict hazard in the direction of non-target species,
- To present information on the method of toxic action,
- To present data scheduled which user risk advantage relationships might assess also,
- To start the exposure leavels into the studies designed in the direction of assess long term effects into the investigational toxicology.

Aluminium has been accepted as a very significant poisonous metal in several temperate freshwater environment other than have received small attention into steamy water environment (Phillips, 1988). Here latest days aluminium have appear to be regard as the main reason in the defeat of fisheries into flexible acidic waters (Schofield in addition to Trojnar, 1980; also Cummins, 1994), various study contain exposed to Aluminium toxicity depend additional on top of pH also speciation (Exley *et al.,* 1994 also Poleo, 1995), have investigate the molecular mechanism concerned into the aluminium toxicity.

During the current study, the absorption - response association was considered by evaluate the middle lethal absorption of aluminium in the direction of *Cyprinus carpio var. communis,* somewhere the reaction is humanity of catch fish.

The whole amount of water retain in a body is maintain at steady point; the main factor scheming the site as well as the amount of liquid in the variety of section are the osmotic services which are maintain by the a variety of solutes such the same as Glucose, Protein in addition to electrolytes (Lehninger, 1982). Fletcher (1970) confirmed so as to the electrolytes of the body fluids contain a variety of function, such as flexible the osmotic pressure furthermore acid-base balance of remains fluids, preserve the permeability of cell, usual neuromuscular tetchiness as well as tissue function.

Na^+ also Cl^- are the main cation along with anion of the extra-cellular solution in that order, where K^+ is the main cation of the intracellular liquid. Sodium furthermore potassium is considered because macro fundamentals also they are necessary to maintain organic life. Circulate levels of sodium contain be use as an pointer of pressure, mainly suitable to the energetic progress of this ion crossways the gill arrangement depending on its absorption in the outside average (Bentley, 1971). Foskett *et al.,* (1983) declared to facilitate blood Cl^- is a significant extra-cellular anion intended for the guideline of cation-anion sense of balance with osmotic stress. These inorganic electrolytes Na^+ and Cl^- contain be often use because a guide of osmoregulation into freshwater teleost less than environment of pressure (Umminger also Gist, 1970).

Maintenance of steady interior ion attention is significant for lively instruction of water arrival plus ion efflux in aquatic organisms (Mayer *et al.,* 1992). Alteration during plasma electrolyte concentrations has proved to be a quite dependable pointer of pressure (Folmar, 1993). Through measure change in plasma ions level, together acute in addition to sub-lethal response to pollutant preserve be successfully monitor; such capacity are simply complete plus the method are economical, moderately responsive as well as present good index of pathological change in ion flexible tissues (Larsson *et al.,* 1985).

Several studies include recognized the progressive acidification of various lakes along with the attendant free of aluminium beginning the earth, which led toward defeat of fish residents (Sayer *et al.,* 1993). Inside fish kill in normal acidic waters, declining fish contain be establish to include unhappy plasma sodium in addition to chloride levels (Rosseland *et al.,* 1994). Acid-base balance inside teleost along with elasmobranchs is a arrangement of HCO_3^-, H^+ and NH_4^+ move among the animal moreover situation, mainly happening next to the gills toward sustain a set position pH inside the body fluids which is necessary for highest metabolic presentation (Wood *et al.,* 1984). Some net manufacture of acidic or alkaline metabolic harvest is objective by equimolar removal beginning the body fluid (Heisler, 1986).

McDonald (1983a) meaningful out to make easy bi-carbonate assume a important place since of its significance in CO_2 secretion along with acid-base balance in addition to its uptake on the gills is equal to the excretion of H^+ which is prohibited through the branchial machinery of acid-base instruction. The review of Brown and Waring (1996) evidently depicts so as to there appear to be a connection among blood acid-base balance in addition to toxicant exposure. A strict refuse inside blood pH in addition to bicarbonate level was experiential in brook trout, *Salvelinus fontinalis* less than acid + A1 introduction by means of Wood *et al.,* (1988, a) and in *Salmo gairdneri* in Heming and Blumhagen (1988). In contrast, still however Acid and aluminium cause cruel decrease in blood pH principles in rainbow trout, *Salmo gairdneri,* it exhibit increased outside concentration of HCO_3^- (Playle *et al.,* 1989).

The previous review of literatures indicates so as to much work have been complete on the poisonous result of metals going on ionic rule in addition to acid-base category into fish. Therefore, in the current work it is considered to learn the authority of aluminium toxicity in impartial pH going on the plasma electrolyte stage in addition to Acid-base balance within *Cyprinus carpio var. communis.* Blood essential control is well recognized for several of the upper vertebrates as analytical gear in human furthermore animal medication. Change in blood parameters is often fast to react ecological or physiological alteration. Further more they are with no trouble quantifiable and give an included calculate of the physiological rank of organisms which live in an exacting dirty place. (Tort also Torres, 1988). Ghazaly (1992) emphasize the require for the learn on haematological in addition to physiological variables since of their relationship among respiration (Haemoglobin contented in addition to connected parameters), defence mechanism (Leucocyte value) in addition to energetics (Metabolite levels).

Stress considerable increase or cut in haematological value of fishes showing to special ecological toxicants have been report by several people (Svobodova, 1982; in addition to Dalela *et al.,* 1991; also Christensen *et al.,* (1972) confirmed to any category of pressure not consequential in disgusting changes in addition to mortality produce certain change in the fish blood individuality. Iwama *et al.,* (1986) optional that change inside the haematology of fish in response to stress agent are indicator of the demanding status in fish in addition to they are helpful in decide when in the direction of start prophylactic events in go fishing society. They can also supply a technique of assess the physiological position of fish.

Alteration inside the haematological parameter such since haemoglobin satisfied, HCT, RBC in addition to WBC count can be induce by minor factors such as imprison and sample (Lowe-Jinde also Nimmi, 1983) since well as constant factors such as contact to infection and ecological contaminants. Haematological technique counting capacity of haematocrit, erythrocyte in addition to leucocyte count have prove important for fishery biologists inside assessing the fitness of fish (Hickey, 1976) as well as in monitor stress response (Soivio also Oikari, 1976). The plan of the current study is to determine the shock of aluminium toxicity in impartial pH going on the haematological parameter of fish *Cyprinus carpio var. communis.*

Histopathological change contain be widely use as biomarkers during the assessment of the physical condition fish expose to contaminats, mutually in the laboratory in addition to field study. Single of the huge advantages of use histopathological biomarkers inside environment monitor is to this kind of biomarkers allow examining particular object organs, include gills, kidney moreover liver to are in charge in support of vital function, such since respiration, flow along with the addition and biotransformation of xenobiotics also the seek(Muller also Triebskorn, 2001).

Haemoglobin (Goss in addition to Wood, 1988), confirmed that opinion of haemoglobin container be use as a directory of anaemia in addition to fluid quantity disturbance. The haemoglobin happy inside fish blood have been important due toward metal experience *viz.,* cadmium in the direction of *Tilapia zillii* (Ghazaly, 1992), chromium in the direction of rainbow trout, *Salmo gairdneri,* (Vander Putte *et al.,* 1982) and in freshwater fish, *Barbus conchonius* (Gill also Pant, 1987), copper to brown bullhead *Ictalurus nebulosus*, (Christensen *et al.,* 1972), brook trout, *Salvelinus fontinalis,* (McKim *et al.,* 1970) along with to *Oreochromis mossambicus* (Cyriac *et al.,* 1989), Zinc toward *Clarius lazera* (Hilmy *et al.,* 1987) moreover mercury in the direction of *Oreochromis mossambicus* (Cyriac *et al.,* 1989).

Haematocrit (HCT), in blood is the relation of the quantity of erythrocytes in the direction of that of the total blood. Reduce inside HCT next to cadmium contact was record inside tilapia, *Oreochromis mossambicus,* (Ruparelia *et al.,* 1990), inside flounders (Larsson *et al.,* 1976), in *Salmo gairdneri* (Haux in addition to Larsson, 1984) also in *Tilapia zillii* (Ghazaly, 1992). Important increase into HCT value in dissimilar fishes showing to chromium (Vander Putte *et al.,* 1982), copper (Cyriac *et al.,* 1989), mercury (Cyriac *et al.,* (1989) in addition to Zinc (Hilmy *et al.,* 1987) was observed.Erythrocyte level was established toward be there miserable inside fishes subjected toward demanding situation (Kumar in addition to Barthwal, 1991). Decrease of RBC count inside fish blood following cadmium into *Carassius auratus* (Houston moreover *et al.*, 1982), in *Tilapia zillii* (Ruparelia, 1990) also lead poisoning inside *Barbus conchonius* (Tewari *et al.,* 1987) have been report.

Variant inside Leucocytes calculation provides an additional sensitive directory of stress than perform changes into erythrocyte profusion (McLeay in addition to Gorden, 1977). Changes inside the leucocyte system manifest in the structure of leucocytosis by means of heterophilia as well as lymphopenia, which are quality leucocyte response inside animals less than pressure. (Sen *et al.,* 1992).

MCHC value declined inside *Clarius batrachus* exposed in the direction of lead (Nath as well as Banerjee, 1995) moreover inside *Channa punctatus* treated among copper (Singh, 1995). Acute exposure toward cadmium shaped apparent reduces inside MCV in addition to MCH into fish, *Tilapia zillii* (Ghazaly, 1992). MCV (Mean cell volume) increase was reported by Nath and Banerjee (1995) in lead nitrate exposed air-breathing fish, *Clarius batrachus.* During the general, MCHC worth was lower in fish payable to aluminium experience in acid water (Vuorinen and Vuorinen, 1991). The previous review of journalism reveals to there is a scarcity of in order on the produce of aluminium toxicity on top of the haematological parameters of fish inside neutral pH. The aspire of the current study is to learn the force of aluminium toxicity inside impartial pH resting on the haematological parameters of fish *Cyprinus carpio var. communis.*

Alexander as well as Ingram (1980) report that the attention of protein into the serum of fish has been uses the same as a pointer of their common condition of physical condition. Proteins display useful versatility in addition to be so, concerned in a collection of organic roles such the same as transport, metabolic manage, contraction in addition to catalysis of substance alteration (Christian *et al.,* 1993). Umminger (1970) reported so as to proteins be the power source through chronic time of pressure.

Greater than before plasma protein point was experiential inside rainbow trout, *Oncorhynchus* mykiss (Goss in addition to Wood, 1988), during *Salvelinus fontinalis* (Wood *et al.,* 1988a) as well as in white fish *Coregonus wartmanni,* (Vuorinen *et al.,* 1990) exposed toward aluminium on low pH. During support of this, Reid *et al.,* (1991) establish important plasma protein happy in fish at what time showing to aluminium inside acidic pH. Vuorinen in addition toward Vuorinen (1991) eminent no change inside plasma protein on pH 4.5 + 333 μg Al in addition to on pH 5.0 with 6.0 with aluminium into fish *Salvelinus fontinalis* in addition to white fish, *Coregonus wartmanni,* in that order. Waring in addition to Brown (1995) also found no changes during plasma protein concentration as revealing brown trout, *Salmo trutta,* in the direction of pH 5.0 + Aluminium.

Plasma glucose has been widely monitored to learn stress inside go fishing (Pickering *et al.,* 1982) confirmed that examine of this significant blood consideration can supply because a pointer of ecological stress. The add to inside blood glucose of catch fish be a usual answer to the majority stressors (Waring and Brown, 1997). Elevation inside blood glucose level has been use by means of Wedemeyer (1972) in addition to Nemcsok with Boross (1982) used for representative metabolic pressure. The elevation inside blood glucose in addition to the resulting hyperglycemic response contain be credited to the let go of hyperglycemic hormones in addition to succeeding mobilisation of glycogen capital because recommended in Mayer *et al.,* (1992).

The previous analysis of journalism reveals to not greatly work have be done scheduled the contact of aluminium toxicity going on biochemical parameters of fish by the side of impartial pH. Therefore the present study is intended to learn the importance of minor stress response (Plasma protein also glucose) in *Cyprinus carpio var. communis,* exposed to aluminium toxicity in neutral pH.

Pollutant into the water ecosystem affects the growth also enzymes inside aquatic organisms (Webb also Brett, 1973). New trend is to fish an aquatic fauna be able to be deliberate since one of the bioindicator organisms, which container survive mainly use to judge the quantity of heavy metals toxicity. The xenobiotics container change Na^+, K^+, and ATPase movement suitable in the way of trouble of power generate metabolic pathway or directly interrelate ion during the enzyme.

Na^+, K^+, and ATPase the stage a middle role in complete body ion regulation by provide power for the progress of Na^+ K^+ crossways the cellular membranes next to an electrochemical grade which be required used for the uptake of metabolites such because glucose in addition to amino acids by means of cells (Evans, 1987). Regeneration of transmembrane potential during nerve excitation, transmembrane movement of Ca^{2+} all through muscle inspiration moreover preservation of osmotic equilibrium inside cells is several of the function credited toward Na^+, K^+, and ATPase (Haya also Waiwood, 1983).

The xenobiotics maintain correct Na^+ K^+ also ATPase movement billed to interruption of control produce metabolic way or in a straight line relate ion between the enzymes (Watson also Beamish, 1980). Gill Na^+, K^+, and ATPase achievement is affected into smolts of *Chinook salmon*, and *Oncorhynchus tshawytscha;* during copper intoxication in normal mechanism water (Beckmann also Zaugg, 1988). In *juvenile rainbow trout, Oncorhynchus mykiss* and in flounder, *Platichthys flesus,* Lauren moreover McDonald (1987) establish depression of Na^+, K^+, and ATPase action by copper in that order.

Aluminium metal inside acid waters have as well been reported by means of Staurnes *et al.,* (1993) in the direction of restrain Na^+, K^+, and ATPase action in *Atlantic salmon, Salmo salar* in addition to and into rainbow trout, *Oncorhynchus mykiss, (*Rosseland *et al.,* 1992). Ganrot (1986) report with the plan of aluminium interferes with basic biochemical function, far above the ground power metabolic also enzyme behavior.

Reserve, of acidic phosphatase action was record inside fish, *Clarias lazera,* in addition to *Tilapia* zillii in zinc contact (Hilmy *et al.,* 1987) as a result of cadmium contact inside *Sarotherodon mossambicus* (Ruparelia *et al.,* 1992) moreover into catfish, *Heteropneustes fossilis,* (Sastry in addition to Subhadra, 1985) also during killifish, *Fundulus heteroclitus* (Jackim *et al.,* 1970). Though increase inside acidic phosphatase action by means of cadmium contact inside bluegill sunfish, *Lepomis macrochirus,* (Versteeg also Giesy, 1986) as well as copper contact into *Oncorhynchus mykiss* (Leland, 1983) be experimental.

Amongst a variety of enzymes study, much notice has been focused on top of alkaline phosphatase (ALPase). By Gill *et al.,* (1992). Alkaline phosphatase is a poly-functional enzyme which hydrolyses a wide class of phosphomonoester substrates into addition in the direction of acts because a transphosphorylase on alkaline pH (Blasco *et al.,* 1993). The enzyme might and play a significant role inside the mineralization of the bones of water animals. Strength of mind of alkaline serum phosphatase is huge significance inside the degree of difference diagnosis of various diseases inside man (Linhardt also Walter, 1965). According toward Boge *et al.,* (1992) inside ecotoxicology, alkaline phosphatase can serve because a good indicator of intoxication since of its sensitivity to metabolic salt. Therefore in the current study the environment along with degree of enzyme answer inside *Cyprinus carpio var. communis* toward aluminium toxicity into dirty waters was investigate.

Increased action of alkaline phosphatase is recorded inside the liver of freshwater go fishing, *Sarotherodon mossambicus* plus in alevins of brook trout, *Salvelinus fontinalis* into cadmium introduction (Christensen *et al.,* 1975). ALP activity is considerably increased as soon as red sea bream, *Chrysophrys major,* be exposed toward zinc (Haya also Waiwood, 1983).

Fish gills, which provide as the main uptake site into fingerlings in support of trace metals, represent the the majority significant target when showing in the direction of elevated level of ambient metals (Newman and Jagoe, 1994). The gills are straight contact by the pollutant medium (water) also include the thinnest epithelium of every one the organs also metals preserve penetrate during the thin epithelia cell. In contrast, toxin concentrations in the other organs (such as the liver or kidneys) depict only the degree of distribution of the toxic compounds (Emilia *et.al.,* 2017). Maximum level of heavy metal increase experimental in the gill compare toward other parts of the body. Since they are in a straight line in contact by water.

Drishya *et al.,* (2016) enclose been report to the histopathological change, the part degeneration of the epithelium of minor gill lamellae, vacuolation, also synthesis, falling apart of gill lamellae, decrease also combination of secondary gill lamellae resultant in decrease of respiratory exterior also vacuolization cause by heavy metal copper into fish).

Starting the common introduction, it is completed to there is a pressing must also requirement to study the contact of hexavalent chromium within aquatic organism into general also fish into particular during order to moderate the metal contamination on the freshwater environment. So the fresh water fingerlings *Clarias batrachus* have been choosen into the current study in the direction of find out the impact of hexavalent chromium on different parameters such because biochemical, histopathological, hematological also enzymological changes into the selected fish fingerlings below the laboratory environment.

Athikesavan *et al*., (2006) experimental that Nickel chlorideon histopathological changes was deliberate in the gill, liver, intestine also kidney of the nickel chloride treat freshwater fingerlings *H. molitrix*.

Cengiz, (2006) report that the histopathological property of deltamethrin on the gill also kidney tissues of *Cyprinus carpio.* The majority common gill changes by every one doses of deltamethrin were desquamation also necrosis. Besides, aneurism into secondary lamellae, exciting of the lamellar epithelium, oedema, and epithelial hyperplasia also fusion of the secondary lamellae is report. Lesions into the kidney tissues of fingerlings showing to deltamethrin be characterize in degeneration into the epithelial cells of renal tubule, pycnotic nuclei within the haematopoietic tissue, dilation of glomerular capillaries, degeneration of glomerulus, intracytoplasmatic vacuoles into epithelial cell of renal tubules by means of hypertrophied cells also reduction of the tubular lumen.

The more than review of text reveals subsequently since to not a great deal job has been concluded taking position the create of aluminium toxicity resting leaving on enzyme movement of fish, *Cyprinus carpio var. communis.* Therefore inside the present learn the normal world by degree of enzyme response contained by *Cyprinus carpio var. communis* in the direction of aluminium toxicity inside neutral waters live current study.

REVIEW OF LITERATURE

Aluminium

Earth external cover include 8% of Aluminium, which is safe within normal property to hide from view its grade at the equal occasion as the mass abundant metal too the 3rd the mass plentiful part, follow oxygen inside adding toward silicon (Mason in addition to Moore, 1982). Interested inside atmosphere, it occurs only in combined figure the similar since an oxide inside bauxite, the major ore plus keens on not easy aluminosilicates such the similar because micas in addition to feldspars.

Aquatic toxic waste of freshwater as a result of metals have suit a grave difficulty since of unsafe effect that far above the ground levels of such substances be able to have on top of water organisms, counting fish. According near Martin (1994), awaiting freshly the majority normal waters contain unimportant amount of aluminium, except for individuals inside a number of volcanic region plus alum spring. While aluminium is an ordinary constituent, its concentration in the majority new waters are regularly extremely low (l mg/ L) (Sadler and Lynam, 1987).

Recruitment of aluminium starting the edaphic toward the water atmosphere is a significant consequence of acid rainfall (Driscoll *et al.,* 1980). Such recruitment, still, is not the only direction for anthropogenic foreword of aluminium toward the hydrosphere (Heming in addition to Blumhagen, 1988). Aluminium compounds inside manufacturing processes, dust pond release, energy conversion effluents, irrigate management works along with dye of textiles might come across their way addicted to hose courses, which exert relentless toxicity toward water biota (Ganrot, 1986).

Howells *et al.,* (1990) report to facilitate small attentiveness of aluminium inside normal stream is partially a result of the association among water pH in addition to the solubility of aluminium raw materials. Birchall *et al.,* (1989) sharp to aluminium toxicity depends resting on the variety of aluminium current (Cationic, impartial or anionic). Aluminium inside answer forms a variety of molecular variety counting the main species Al^{3+}, $AlOH^{2+}$, $Al(OH)_2^+$ in addition to $Al(OH)_4^-$ depending mainly happening pH of stream. The hydrolysis harvest, Al $(OH)^{2+}$ plus Al $(OH)_2^+$

make up the leading parts of the inert dissolve aluminium which is the majority poisonous on a pH point 5.0 to 6.0.

Aluminium hydroxides comprise a considerable piece of the dissolve aluminium among pH 5.0 and 6.0 while toxicity is most brutal in addition to a fast rise inside pH considerably increase the toxicity of aluminium toward fish (Poleo *et al.,* 1994). So, less than such circumstances, it have been planned that the development of constant aluminium polymerization be the main foundation of delicate aluminium toxicity just before fish (Poleo *et al.,* 1994) in addition to not because before optional, the monomeric A1-hydroxides themselves (Sadler in addition to Lynam, 1987). Thus, the severe toxicity of aluminium just before fish is a substance of reaction kinetics, wherever the quantity of continuing aluminium polymerization is the the majority significant reason (Lydersen *et al.,* 1991).

Burrows (1977) as well as Ganrot (1986), as review the position of aluminium in biology plus chemistry explain that aluminium is there in very little amounts inside alive organisms in addition to inside rejection container have it be shown in the direction of have a specific organic purpose; aluminium have generally be regarded because virtually in nature inert in addition to the significance shown intended for its biochemistry, metabolism in addition to toxicology have been partial. However, throughout fresh years, a growing number of poisonous properties through aluminium contain been establish (Atland, 1998).

The current thoughtful of the produce of Aluminium toxicity within normal waters be based mainly on a complimentary ion movement form derived on substance balance (Campbell, 1988). Incident the contrary, Exley *et al.,* (1996) recognized that the power of society of momentary aluminium hydroxide polymers among gill mucous was the shaping part for severe toxicity moderately than the free of charge aluminium ions spring at the gill outside in non-steady state individuality of the normal atmosphere (Exley and Birchall, 1992).

The main target organs for toxicants are surface sensory organs plus the gills in fish (Galvez *et al.,* 1998). Precipitate aluminium complex container gets on your nerves the gill in addition to reason irritation, edema, bulge, as well as from time to time irradiation of the minor lamella; initiate a sequence of opposition mechanism

(Goossernaerts *et al.,* 1988). Possible sites of aluminium communication at the lamellar epithelium be able to explain altering membrane variability in addition to permeability individuality, as fine as intracellular uptake of aluminium with minor effects going on enzyme systems, intracellular calcium, with energy metabolism; the net product of the belongings on these process is inside the majority gear is osmoregulatory not a success (Waring as well as Brown, 1995).

Freeman as well as Everhart (1977), reported to publicity of go fishing to Aluminium (5.2 ppm) at pH 7.0 cause reserve of feed after 4 hours of contact. Activity is summary after 4 being as well as hyperplasia of gills are experiential behind 7 days. Humanity occur at a usual rate all through the examination episode of 45 days in addition toward the survivors be really emaciated averaging single 24.1 percent of the manage influence since opposite to 91.8 percent of the manage.

Few study to calculate the toxicity of aluminium effluents be conduct on *Daphnia magna* (Biesinger also Christensen, 1972) lying on midge larva, *Tanytarsus dissimilis* (Lamb also Bailey, 1981), *Ceriodaphnia dubia, Pimephales promelas,* (Hall also Hall, 1989) also *rainbow trout, Salmo gairdneri,* largemouth bass, *Micropterus salmoides* along with the patterned salamander, *Ambystoma opacum,* (Birge *et al.,* 1980) by the side of impartial also circum on the fence pH.

Since the previous assessment of journalism, it is understood that not greatly work have been complete on aluminium induce physiological moreover biochemical change inside fish on impartial pH. Therefore, the current work is deliberate to learning the physiological as well as biochemical responses inside fish, *Cyprinus carpio var. communis,* showing toward aluminium toxicity inside impartial soft stream through reference toward the subsequent aspect.

OBJECTIVES OF THE PRESENT STUDY

The majority main steady is the LC_{50} (Median lethal concentration), which be the hypothetical value, cause 50% humanity in the populace organism study. The LC_{50} worth is resolute following 24 hours or 48 hours of experience in test of acute toxicity furthermore into some luggage behind 96 hours (Ramade, 1987).

Toxicant

Metal (Aluminium)

According in the direction of aluminium is an main metal by means of the possible to restore zinc as well as tin inside various finish user industry; aluminium chloride is use at the same time as a pigment inside the petroleum along with natural chemical industry because a mechanism; aluminium sulphate be used inside the pulp with paper industry along with inside treatment of clean water supply somewhere it act because a coagulant. Aluminium sulphate (anhydrous form) use in support of the current study was obtained beginning New India Chemical Enterprises, Chennai-600 017 Tamilnadu also India.

- Determination of Median Lethal Concentration
- Biochemical mechanism (Protein in addition to Glucose)
- Preparation of Sample (Enzyme)
- Estimation of Adenosine Tri-phosphatase (ATPase)
- Sodium in addition to Potassium ATPase
- Estimation of Acid Phosphatase furthermore Alkaline Phosphatase
- Haematological Studies
- Histological study of gills.

Into the present study the natural history also degree of enzyme answer in *Cyprinus carpio var. communis* in the direction of aluminium toxicity inside infected waters be present investigate.

MATERIAL AND METHODS

MATERIALS

Specimen of scale carp, *Cyprinus carpio var.* communis, (Linnaeus) an exotic fish belong in the direction of the family Cyprinidae be chosen for the current study base on the subsequent criterion.

- The family Cyprinidae is healthy represent among the piscine residents of the fresh waters of India.
- With pleasure presented all through the year inside plentiful capacity, commercially significant in addition to circulated all through the earth.
- It is responsive to ecological factor less than deliberation.
- Breed into confine waters.
- Voraciously omnivorous, professionally convert the foodstuff ingested interested inside tissue, grow especially quick in addition to horizontal toward reproduction feed.
- Understanding to broad fluctuations of high temperature in addition to opposed to disease.
- Hardy natural world for management also transportation.
- Appropriate creature in support of bioassay test.

Experimental Design

- 24 hours' Median lethal concentration of aluminium sulphate toward fish, Fresh water fish in *Cyprinus carpio var. communis.*
- Electrolyte disturbances furthermore Acid-base imbalance inside the blood as well as plasma of fish all through acute in addition to sub-lethal experience toward aluminium because an indicator of ion-regulatory breakdown.

- Change into the haematological individuality of the blood of fish treat through acute in addition to sub-lethal attention of aluminium sulphate toward charge the solution volume in addition to oxygen uptake instability.
- Biochemical moreover enzymatic alteration inside the plasma also gills of fish, correspondingly, all through acute also sub-lethal aluminium exposure toward appreciate cellular also physiological dysfunction.

Recruitment of Fish used for conduct experiment

Well specimen of *Cyprinus carpio var. communis* be chosen as well as transported to laboratory into polythylene bag contain oxygenated water since Tamilnadu Fisheries improvement Corporation Limited, Aliyar fish farm, Aliyar village, Pollachi near, Coimbatore district, Tamil Nadu, India. Fish of the equal age also size since the equal brood stock be present collect.

Fish stock is maintained inside an uncontaminated clean large rectangular tank before clean with water also clean by means of potassium permanganate. They are acclimatizing in the direction of laboratory condition used for a fortnight, previous to organism use for experiment. Media is changed commonly to keep away from fungal enlargement in addition to infectivity by metabolites. These stocks are fine aerated by means of aerators.

All through the adaptation stage, fish stock is feed *ad libitum* by resources of groundnut smear with oil cake with rice bran powder into the relation of 1:2. Feed was closed two days previous near the commencing of experiment to keep the investigational natural world more or less inside the similar state of metabolic condition. All through adaptation, the fish collection was maintained at normal photoperiod along with ambient high temperature. Since this collection fish by way of a common length of 12-15 c.m also weight 32-38 g is segregate in addition to transfer to clean rectangular glass aquarium tank (75 X 35 X 37 cm) of 150 Litter water capacities.

PLATE - I
Systematic position of Common *carp, Cyprinus carpio var. communis*

Classification

Phylum	-	Chordata
Group	-	Vertebrata
Series	-	Pisces
Class	-	Teleostomi
Sub-class	-	Actinoptergii
Order	-	Cypriniformes
Division	-	Cyprini
Family	-	Cyprinidae
Sub-family	-	Cyprininae
Genus	-	*Cyprinus*
Species	-	*carpio*
Variety	-	*communis*

Toxicant

Metal

According in the direction of Howells *et al.,* (1990) aluminium is an significant metal with the possible to change zinc and tin into different finish user industry; aluminium chloride is use as a color in the petroleum furthermore natural chemical industries because a catalyst; aluminium sulphate is use into the pulp also paper industry with in management of clean water supply where it act as a coagulant.

Aluminium sulphate (Anhydrous variety) use for the current study was obtained from New India Chemical Enterprises, Chennai, Tamilnadu and India. The physical in addition to chemical property of aluminium sulphate (No. Al1429) live as follow:

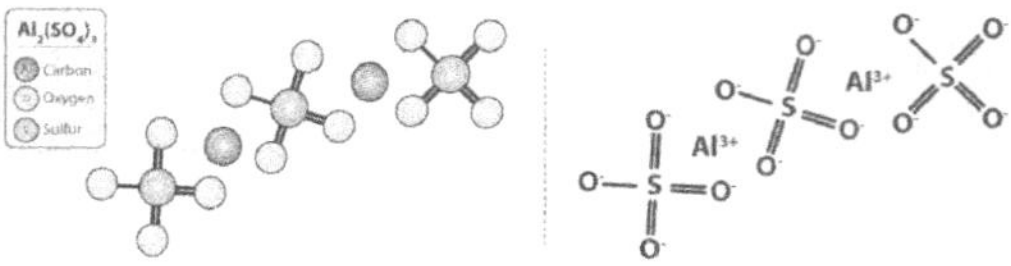

Physical property

A. Formula - $Al_2 (SO_4)_3. 16H_2O$
B. Molecular control - 342.15
C. Colour - White
D. Purity - 97% as $Al_2 (SO_4)_3$

Chemical possessions

According in the direction of Driscoll furthermore Schecher (1988), aluminium into solution is amphoteric also can form mutually organic also inorganic complexes, treatment to polymerize; it be a group III component establish in solution only inside the trivalent state; though it is metallic, it exhibit noticeable covalent tendencies in addition to thus form relatively steady complexes through a variety of lifeless also organic material.

WATER QUALITY

Tap water free beginning chlorine is use for the current study. The hydro organic features such like Temperature, pH, dissolve oxygen, Total alkalinity, Salinity in addition to Total hardness, be estimated in support of each set of research as these factor contain a important power happening the biodegradability along with toxicity of pollutant.

Study of Water Chemistry

Water temperature, pH, Alkalinity, Hardness, Calcium as well as Dissolved oxygen be monitor every day both into control moreover investigational tanks. Water temperature was deliberate by a thermometer. pH values are resolute by a fountain pen type pH meter (PH Scan 1, Eutech Cybernetics PTE Ltd). Dissolve oxygen satisfied was predictable by means of Winkler's technique by starch indicator. Total alkalinity is strong-minded by methyl orange because a pointer. Total hardness is projected use Erichrome Black - T indicator in addition to calcium stage be resolute use murexide pointer. Other parameters such since salinity, fluorine, silicate in addition to dissolve natural carbon are projected weekly. The exceeding physico-chemical study of water use in the current experiments is carried out as per Apha *et al.,* (1985). The investigative statistics for an only set of experiment for the over parameters are given in Table 1, because the value various negligibly for the waters use for the additional set of experiment.

Table 1. Physico-chemical facial appearance of water make use of in maintain of the present study.

S.No	Physico-Chemical Features	Values
01.	Temperature	26.0 ± 1.0°C
02.	PH	7.7 ± 1 units
03.	Dissolved oxygen	7.3 ± 0.02 mg/L
04.	Total alkalinity	36.0 ± 5 mg/L
05.	Salinity	0.23 ± 0.1 ppt
06.	Total hardness	18.0 ± 0.5 mg/L
07.	Calcium	4.25 ± 0.5 mg/L
08.	Fluoride	Nil
09.	Silicate	0.001 ppm
10.	Dissolved organic carbon	Nil

Value be means ± S.E. of five individual observations

Two type of bioassay test *viz.,* Continuous flow (Mount in addition to Warner, 1965) also Static (APHA, 1971), the static bioassay method be select allow for the limits of the laboratory services.

Toxicant research in addition to contact

Aluminium collection solution were ready by dissolve 1 gm of aluminium sulphate ($Al_2(SO_4)_3 . 16H_2O$) in 1 litter of tap water in addition to store into stoppered acid wash Erlenmeyer flasks. Tests were conduct into previously clean, sundried circular plastic tubs of 25 litters' ability. Beginning toxicity test be accepted out to discover absent Median lethal tolerance maximum of fish in the direction of aluminium sulphate used for 24 hours. Use for determining the LC_{50} absorption, plastic tubs investment 10 litters of water be use moreover appropriate concentration of aluminium sulphate be mix to the test water into the tubs. In each tub 10 fishes are present introduced. A control tub is too maintained with 10 fishes in 10 litter of water.

Determination of Median Lethal Concentration

In the end of 24 hours the survival / mortality of fish into the control as well as experimental tubs are recorded. The deliberation at which 50% mortality of fish occur after 24 hours was in use as the Median lethal concentration (LC_{50}) for 24 hours. The LC_{50} absorption for 24 hours was calculate by the probit - study method of Finney, (1978). Dead fish were indifferent immediately from the experimental tubs. Fish is measured deceased when they missing their equilibrium, balanced belly up, become stationary, cessation of ventilatory in addition to mouth movements also inability in the direction of respond near any incentive.

Calculation of Regression line

Critical concentration or vulnerability of an individual to any toxicant can be probable with satisfactory accuracy since a probit/ log absorption graph (Busvine, 1971). The two variables are plot on a plain document, or the original data (Percentage mortality as well as concentration) preserve be plotted resting on a logarithmic possibility document. A directly line is integral by eye furthermore the significant concentrations be resolute by examination. Values strong-minded graphically are often remarkably close up to calculate results, although they present no specific information going on limits of truth. Base on the more than method, the degeneration line was calculated in the current reading.

Experimental Protocol

Two investigational progressions be present *viz.,* Acute through sub-lethal toxicity learning. In every one study one control also five replicates of management groups are maintain.

Acute Toxicity Studies

In favor of acute toxicity studies, strong fish be present taken beginning the stock along with be maintain in the glass tank. Subsequently two days proceeding to experiment also during the experimental stage feeding be discontinued. For sensitive aluminium toxicity study, five spherical plastic tubs among 25 Litters ability were taken also each are filled through 20 Litters of water. The pH of the water is maintained at 7.7 ± 1. Then LC_{50} 24 hours' attention of aluminium sulphate (44.90 ppm) is extra to each one tub furthermore 20 fishes be introduced into each tub also mixed well. In favor of control groups, a spherical synthetic tub of 25 Litters capacity was in use with 20 Litters of water of common pH 7.7 ± 1. Subsequently 20 strong fishes are selected from the collection also introduced interested in the tub in addition to no toxicant be supplementary.

Sampling

By the side of the conclusion of 24 hours, be alive fish from control along with experimental tubs is taken in addition to sacrificed used for electrolyte studies. A minimum of 20 fishes per management, 5 replicates per treatment also 20 fishes per duplicate be use for acute study. Care is in use to keep away from stress all through sample.

Sub-lethal Toxicity Studies

A glass container of 120 Litters capability was taken furthermore full with 100 litters of stream. The pH of the control in addition to new work is maintained at 7.7 ± 1. 1/10 of the $LC_{50;}$ 24 hours' attention of aluminium sulphate (4.49 ppm) be in use because the sub-lethal absorption (Sprague, 1971) with mixed by the water inside the tank. One hundred strong fishes be select from the stock along with be introduced interested in the water tank, five comparable replicates be maintain. Water into the tank is changed every day with renewed by the toxic resolution. A separate control is maintained by stock 100 fishes into a glass tank exclusive of addition toxicant. Fish is fed *ad libitum* each day. Water was every day changed.

Sampling Frequency

Experiments are conducted used for a stage of 35 days among weekly sample regularity. No humanity was experimental during the investigational period. Winning completion of the predetermined revelation periods of 7, 14, 21, 28 as well as 35 days, 20 fishes be taken out in addition to sacrifice without freezing for additional investigation. Before removal of fish by the side of various interval of time, the quantity of the investigational along with control media is adjusted to continue a constant solidity of fish per component capacity of water.

Collection of Blood

Fish starting experimental in addition to control tubs is in use out used for sub-lethal toxicity study. Blood is drawn in cardiac puncture use plastic non-refundable syringe fixed by 26-gauge needle. The syringe with needle be prechilled furthermore moisten among heparin (Beparine^R heparin sodium, IP 1000 IU ml^{-1} resulting beginning beef intestinal mucosa contain 0.15% w/v chlorocresol IP preservative), an anti-coagulant manufactured by Biological E Limited, Hyderabad, India.The collected blood be then decant interested in ice cold heparinized plastic vials be used. The blood sample starting every part of the fish of their individual treatments is pooled for the study of dissimilar parameters. Blood sample be centrifuged at 10,000 rpm used for 10 minutes. This left a comprehensible yellow solution, the plasma, which be used used for the resolve of Protein, glucose, Na^+, K^+, ATPase, acid, alkaline phosphatases in addition to chloride. Total blood collect in vials be used in support of Na^+, K^+, pH, Bi-carbonate, Haemoglobin, Haematocrit, RBC also WBC study.

Estimation of Electrolytes

Sodium, potassium in addition to chloride ion substance in the blood furthermore plasma of fish, in that order, be analysed to learning the electrolyte commotion by Acute as well as sub-lethal aluminium experience in *Cyprinus carpio var. communis.*

Estimation of Sodium

Sodium is estimated with flameless ion-selective electrode technique use KNA2 Sodium-Potassium Analyzer.

Principle

The Radiometers KNA2 Sodium-Potassium Analyzer is a computerized scheme for simultaneous purpose of potassium in addition to sodium use K and Na Ion-selective electrodes (ISE). The ISE straight actions the absorption of potassium (K^+) along with sodium (Na^+) into whole blood later than manual dilution in addition to and it events the movement of the ions in the aqueous phase by resources of an F2211 Na Electrode (The displayed charge is articulated as attention of sodium *i.e.,* in m.mol/ L). The Sealed Calomel Reference Electrode (K 1702) be used when a suggestion electrode for together the K in addition to the Na pointer electrodes.

In support of rinsing along with zero-point calibration of the K in addition to Na electrodes, the S3641 rinse in addition to calibrating answer for normal K^+/ Na^+, is use also S2422 calibrating result is use for high K^+, low Na^+ used for compassion calibration of the K and Na electrodes. The S4512 new salt-bridge answer is future as a salt bridge involving the orientation electrode moreover the K also Na electrodes. The measure unit with all the electrodes is thermostatted at 37 ± 0.1°C.

Procedure

The Sodium-Potassium Analyzer KNA2 is setup by means of mode key system to get P form to approximate whole blood sodium. Subsequently the heparinized blood sample was aspirated during the inlet stub in use a catheter tube as well as immediately urgent the aspirate button, cultivate the eventful lamp glow. Then the tube be removed also the inlet flap be turned off. The total sample needed in support of measurement is 125 μl. The sodium activity into the aqueous phase be calculated by ISE (F2211 Na SELECTRODE) within 47 seconds in addition to the result be displayed since the attention of sodium articulated in m.mol/ L.

Estimation of Potassium

Potassium is estimated use KNA2 Sodium-Potassium Analyzer (Radiometer Copenhagen, Denmark) by means of flameless ion-selective electrode technique.

Principle

The principle used for potassium estimation be similar toward that of sodium although, the activity of potassium in the aqueous phase be measured by means of means of an F2321 K SELECTRODE (The displayed value is articulated since concentration *i.e.,* m.mol/L).

The S3641 rinse- furthermore calibrating answer used for standard K^+/ Na^+ study is use for rinsing also zero-point calibration of the K along with Na electrodes. The S2422 calibrating solution in support of high K^+, low Na^+ is use for feeling calibration of the K also Na electrodes. The S4512 new salt-bridge solution is future while a salt- bridge involving the reference electrode in addition to the K also Na electrodes. The measure element by means of all the electrodes is thermostatted by $37 \pm 0.1°C$.

Procedure

The form button system be setup toward obtain P mode to study complete blood potassium in the KNA2 sodium potassium analyzer. The sample (Heparinized blood) is aspirated during the inlet stub use a catheter tube also urgent the aspirate button immediately cultivate the busy lamp glow. Afte r that the tube be removed in addition to the inlet flap is closed. The total sample wanted for dimension is 125 µl. The potassium action in the aqueous phase was deliberate by means of ISE (F2321 SELECTRODE®) within 47 seconds moreover the result be displayed because the attention of potassium articulated in m.mol/L.

Estimation of Chloride

Chloride was estimated through modified method of Schoenfeld in adding to Lewellen (1964) moreover Levinson (1979) use Stangen (ECA 3) electrolyte diagnostic reagent kit artificial plus supplied in Stangen Immunodiagnostics, Hyderabad, India.

Reagents

Reagent I: Chloride reagent

Reagent II: Standard chloride reagent (100 m.mol/L).

Procedure

Four test tubes labelled Blank, standard, control in addition to test were taken plus to each tube 2000 μl of chloride reagent (Reagent I) be added. Afterward 10 μl of deionised water be added to blank as well as to standard, 10 μl of standard chloride reagent (Reagent II) be added. Also, 10 μl of plasma beginning control in addition to experimental groups be added to control also test tubes, correspondingly. The contents in the tubes be mixed well as well as allowed to stand next to room temperature for 2 minutes in addition to the absorbance of standard, control also test was measured at 505 (Hg 546 nm) against blank by Photometer 4010 System.

Calculation

$$\text{Chloride in m.mol/L} = \frac{\text{Absorbance of Sample}}{\text{Absorbance of Standard}} \times 100$$

Concentration of standard chloride = 100 m.mol/L

Statistical Analysis

Statistics be analysed statistically in addition to the mean worth of five individual explanation be taken for each parameter. The importance of sample principal involving control in addition toward treated fish is tested by means of students't' test (Campbell, 1988). The standard error used for the sample mean was calculated as well as they are given in the suitable tables. The importance of sample mean involving control along with the treatment were analysed by analysis of variance.

Haematological Studies

To assess the haematological profile of the control and treated fish, haemoglobin, haematocrit levels, RBC and WBC counts were measured in the whole blood of *Cyprinus carpio var. communis.* Erythrocyte indices of fish *viz.,* MCHC (Mean Cell Haemoglobin concentration), MCH (Mean Cell Haemoglobin) and MCV (Mean Cell Volume) were also calculated.

Haemoglobin

Haemoglobin content was analysed by means of using ABL 300 acid base laboratory.

Principle

The ABL 300 Acid Base laboratory be an automated in addition to computerized micro blood gas as well as acid-base balance analyzer which measures Hb in haemoglobin photometer in addition to pH, PCO_2, PO_2 by their respective electrodes going on blood samples as well as calculate HCO_3, TCO_2, ABE (Actual Base Excess), SBE (Standard Base Excess), SBC (Standard bi-carbonate), SAT (O_2 saturation), O_2CT (Oxygen content) starting the considered values.

The haemoglobin photometers provide a monochromatic light source by means of a 505 nm green interference filter. A transmittance principle is use with the light passing through 2 lenses, the filter in addition to an additional lens previous to attainment the cuvette. The haemoglobin capacity is complete into the first 10 seconds of the measure programme.While a calibration occur, the print out reads Hb values. The Hb zero-point calibration is performed on the red calibrate answer ever since this have no haemoglobin contented.

The haemoglobin content is derived starting Beers equation.

$$cHB = K1 (\log I_0 - \log I_t)$$

Where,

cHB =Haemoglobin content

K_1 = Constant

I_0 = Intensity of incident light

I_t = Intensity of transmitted light

Procedure

The ABL 300 Acid-Base Laboratory is setup to obtain the ready lamp (Green) to glow to analyse haemoglobin content into the sample.The heparinised blood sample was then aspirated from end to end the inlet stub using a catheter tube in addition to pressing the aspirate button immediately. The sample be aspirated till the sample lamp (Yellow) glowed. Then, the tube was removed in adding together to the inlet flap was tightly closed. The total sample needed for quantity is 85 µl. The haemoglobin content was deliberate within 10 seconds of the measure programme *i.e.,* 71 seconds in addition to the result was display the same as the concentration of haemoglobin in g/dl.

Haematocrit (Packed Cell Volume)

Haematocrit be alive estimated inside micro-haematocrit (Capillary) method into sodium heparinised micro-haematocrit as explain among Nelson also Morris (1989), by means of RM 12C micro centrifuge (Remi Centrifuge, Remi Motors, Bombay, India) furthermore a micro-haematocrit reader.

Principle

While anti-coagulated total blood is centrifuged with a constant speed, Erythrocytes (RBC), which is heavier than White cells (WBC), platelets furthermore plasma, settled on top of the bottom. This Red cell column is called haematocrit or packed red cell volume which is expressed because fraction of the complete blood (Level of plasma). Throughout micro-haematocri method, the anti-coagulated blood is alive centrifuged within a sealed capillary tube because fine as the volume of packed red cells in addition to percentage of the total blood (Level of plasma) be resolute by an exacting haematocrit booklover.

Procedure

Two sodium heparinised capillary micro-haematocrit tubes each regarding 7 cm long during a uniform bore of regarding 1 mm be taken. In that case, the blood first control into totaling to new groups kept in their respective vials is filled up to 5 cm in capillary desirability. The sucking finish of each tube was sealed by means of modeling clay in addition to the filled tubes be placed inside the radial grooves of the micro-haematocrit centrifuge start with the sealed end away from the centre in adding to speed by 10,000 rpm used for 5 minutes by RM 12 C micro centrifuge. The tubes be then taken in addition to the length of the total column include plasma along with that of the red cell column alone be measured into a millimeter rule of the micro-haematocrit reader. The concentration of the red cells was taken since the haematocrit charge which is expressed into proportion.

RBC Count

Erythrocytes were counted via the method of Rusia in addition to Sood (1992), by means of haemocytometer.

Principle

The blood specimen is diluted by means of red cell diluting fluid. It destroys the white cells, but allows the red cells toward be count inside a known volume of fluid. Finally the number of cells during undiluted blood is calculating in addition to reported as the number of cells per cubic millimetre of complete blood.

Procedure

A clean RBC pipette was taken in addition to blood was drawn up in the direction of the 0.5 mark. The tip of the pipette be wiped clean down by dipped vertically interested into the red cell diluting fluid, which was after that sucked up to mark 101. Then, the tip of the pipette was stopped by means of a finger in addition to the contents be mixed thoroughly through shaking the pipette on right angles toward its long axis. The red bead into the bulb helps into carry of proper mixing. The ruled part of the haemocytometer is focused clearly. In that case the first drop of the diluted blood inside the pipette was discarded. Through holding the pipette on an angle of 45°, the diluted blood be blown, interested in the counting chamber. The blood was drawn keen on the chamber in capillary action. Then, the counting chamber is left for 3 minutes to allow the cells to settle down.

Counting

The slide was primary examined under low power in addition to in that case under high power magnification. The counting chamber of the haemocytometer has a central heavy ruled area of 1 sq. mm. This central area is the RBC counting chamber. It is alienated in to 25 squares and each square is sub-divided into 16 small squares. For the erythrocytes count, the cells lying within as well as those touching the right moreover higher margin of the four corner squares as well as the central square (80 Small squares) were count. The total number of erythrocytes per cubic millimetre of overall blood was next calculating.

Calculation

$$\text{Number of erythrocytes (Million/cu.mm)} = \frac{\text{Number of erythrocytes counted x Dilution of blood}}{\text{Area counted x Depth of fluid}}$$

Dilution = 200

Area counted = 5 x 0.04 = 0.2 sq.mm

Depth of fluid = 0.1mm

Leucocytes Count

Leucocytes are count through the method of Rusia as well as Sood (1992), by haemocytometer.

Principle

Blood is diluted among acid solution which removes the red cells inside haemolysis furthermore accentuates the nuclei of the WBC cells. Counting is complete with a microscope below low power along with knowing the volume of fluid examined in addition to the strength of the blood, the number of WBC cells per cubic millimetre in undiluted total blood is calculated.

Procedure

Blood was drawn up and about to the 0.5 mark by a clean WBC pipette. behind that, the pipette was immediately kept in a watch glass containing WBC diluting fluid in addition to it was drawn up to mark 101, taking care to no air bubble was integrated. The contents were miscellaneous fine via, rotating the pipette involving the palms of the hands. The white bead in the pipette helps used for proper mixing of blood by the diluting fluid. The diluted blood was allowed toward situate since such for 3 minutes in maintain of haemolysis of red cells to happen. Again the contents be mix by revolving the pipette. After disposal the primary few drops of dilute blood, the counting chamber of the haemocytometer was charged with the fluid, making certain to no air bubbles be trapped flanked by the cover slip as well as the compartment. The cells were allowed toward settle down used for a minute.

Counting

In support of the counting of leucocytes, the slide was examined under low power magnification of microscope. The Neubaeur's counting chamber is divided keen on two counting area which be ruled. Every one counting chamber is alienated addicted to a total ruled area of 9 sq. mm. The area of every one square is 1 sq.mm. The 1sq.mm area by the 4 corners of the slide was use for the counting of leucocytes. The cells declining inside the four corner squares be count in adding to the whole number of leucocytes per cubic millimetre of complete blood was calculate.

Calculation

$$\text{Number of Leucocytes (Thousands/cu.Mm)} = \frac{\text{Number of Leucocytes counted x Dilution of blood}}{\text{Area counted x Depth of fluid}}$$

Dilution = 20

Area counted = 4 x 1 = 4 sq.mm

Depth of fluid = 0.1 mm

1. Mean Cell Haemoglobin Concentration (MCHC)

The MCHC is the normal concentration of haemoglobin into a particular volume of packed red cells. It is calculated starting the haemoglobin concentration in addition to the haematocrit.

$$\text{MCHC (g/dl)} = \frac{\text{Hb (g / dl)}}{\text{HCT (\%)}} \times 100$$

2. Mean Cell Haemoglobin (MCH)

The MCH is the content (Weight) of haemoglobin of the average red cell. It is calculated from the haemoglobin concentration in addition to the RBC cell count.

$$\text{MCH (Picograms)} = \frac{\text{Hb (g /dl) x 10}}{\text{RBC (Million/ cu.mm)} \times 10^6}$$

3. Mean Cell Volume (MCV)

The MCV is the average volume of Red cells as healthy as is calculate starts the haematocrit (HCT) along with the red cell count (RBC).

$$\text{MCV (Cubic micra)} = \frac{\text{HCT (\%) x 10}}{\text{RBC (Millions/ cu. mm)} \times 10^6}$$

Plasma samples from fish treated with Aluminium sulphate next to acute in addition to chronic levels is taken as per the procedure outlined contained for Protein and glucose estimation.

PROTEIN

Total protein concentration be analysed following Biuret method because described by Strickland *et al.,* (1961), Henry moreover Winkelman (1974) by Autopak protein diagnostic reagent kit manufactured in addition to supplied by Miles India Ltd., Baroda, India .

Reagents

Reagent I : Biuret reagent,

Reagent IA: Surfactant reagent,

Reagent II: Standard albumin reagent (7 g/dL).

Solution A

Working solutions

In the direction of the absolute volume of biuret reagent (Reagent I), 81.00 ml of distill water was supplementary. After that complete quantity of surfactant reagent (Reagent IA) be added in addition to mixed smoothly to let alone foam.

Procedure

The Technicon RA-500 system, a computer proscribed, random right of entry scientific chemistry analyser organization be system for the study of whole plasma protein absorption into the sample. Previous to first the experiment, the calibration factor be establish absent. This worth was resolute *via,* a calibrator assess according in the direction of the lessons into the chemistry program statistics sheet used for *in vitro* capacity of whole protein inside the Technicon technique manual of Technicon RA-500 system. One time if the calibration as well as program setting be complete the instrument functions according toward the fed value pending the next calibration.

Once setting the instrument, the working solution (Solution A) be located in the reagent tray with the sample beginning control in adding to investigational groups be placed inside appropriate groove of the model plate. After that, 2 purpose be operated follow by i.d, number (According to our expediency therefore as toward identify the model) inside order to download the job list by give a exacting analyze code. While the work list have been cultured in addition to operate command have been agreed, the Technicon RA-500 system operate mechanically as well as 375 μl of solution A for every one test moreover 7.5 ml of test from the individual vials be added to the response tray contain permanent cuvettes. The absorbance of the model was deliberate moreover calculated as given below in the computing component in addition to the total protein attention in g/dl be report as printed matter inside 2 summary.

Calculation

Total protein attention in g/dl is calculated since follows:

$$\text{Reported result in g/dl} = \frac{\text{Assay result x Unit factor - Intercept}}{\text{Slope}}$$

Where,

Assay result = Calibration Factor x Assay Absorbance.

GLUCOSE

Correct enzymatic Glucose is probable according to the technique of Trinder (1969), Bergmayer (1974) in addition to Young *et al.,* (1975) enzymatically by means of glucose oxidase/ peroxidase method with Autozyme glucose diagnostic reagent kit artificial by Accurex Biomedical, Bombay India.

Reagents

Reagent I : Enzyme vial,

Reagent II : Diluent reagent,

Reagent III : Standard glucose reagent (100 mg/dL).

Solution A

Working Solutions

Toward the working solution bottle the contented of enzyme vials be transfer. Subsequently the diluents reagent is added in addition to mix quietly by swirling or inversion. Worry is taken to circumvent shaky strongly.

Components in addition to Concentration of Working Solutions

Phosphate buffer, pH 7.0	= 170.00 mMol/L
Glucose oxidase	= 5000.00 IU/L
Peroxidase	= 3000.00 IU/L
4 amino antipyrine	= 0.28 mmol/L
Phenol	= 16.00 mMol/L

Stabilizers and inactive ingredients.

Procedure

The Technicon RA-500 system is a computer proscribed random right to use clinical chemistry analyser organism. It is setup for the investigation of correct enzymatic glucose into the plasma. Used for this, primary the calibration reason was determined. This value is determined *via,* a calibrator assay according toward the lessons into the chemistry program data sheet used for *in vitro* quantity of enzymatic glucose into the Technicon way manual of Technicon RA-500 System. The calibration as well as plan setting is complete for one time previous to running the experiment.

The plasma beginning control moreover experimental groups are placed inside appropriate groove of the model tray. Subsequently the 2 functions as well as i.d, number (particular according to our expediency so as toward identify the model) were operate to enter occupation list in addition to download the occupation list with charitable a exacting assay rules. At present the Technicon RA-500 System operate automatically as well as 350 μl of solution A used for every one sample and 7.5 μl of plasma beginning the particular vials be added individually to the personal fixed cuvettes during the response plate. The inside is incubated designed for 30 seconds along with the absorbance of the sample be deliberate independently. The computing component planned the accurate enzymatic glucose stage within mg/ dl as particular below, also printed out within 1 minute.

Calculation

Correct enzymatic glucose within mg/ dl was calculated as follow.

$$\text{Reported result within mg/dl} = \frac{\text{Assay result x Unit factor - Intercept}}{\text{Slope}}$$

Where,

Assay result = Calibration Factors x Assay Absorbance.

Na^+, K^+, ATPase action in Gills, Plasma, Acid, in addition to alkaline phosphatase activity inside plasma of control moreover experimental fishes be study.

Preparation of Samples (Enzyme)

Blood be collected beginning live fish of control moreover experimental groups in support of enzyme examine. The blood be drawn starting the heart section by cardiac puncture by a syringe, pre-rinsed among the anti-coagulant heparin as well as it be transferred interested in little vials. The model was centrifuged by 9000 rpm for 5 minutes. This left a apparent yellow fluid, the plasma, which was use for the resolve of Na^+, K^+, ATPase, Acid phosphatase in addition in the direction of Alkaline phosphatase.

GILLS

The gills be separated starting the control also experimental fish, plus 200 mg tissue starting each was weight. They be homogenised with 1 ml of 0.1 M. Tris-HCl buffer (pH 7.7) use a Teflon homogeniser also subsequently centrifuged at 10,000 rpm at 4^0C for 15 minutes. The supernatant was use for Na^+, K^+, in addition to ATPase activity study in gills.

Estimation of Adenosine Tri-phosphatase (ATPase)

Sodium and Potassium ATPase

The specific activities of Na^+, K^+, also ATPase were assayed following the method of Shiosaka *et al.,* (1971).

Reagents

(i). Tris-HCl Buffer Solution (0.1 M; pH 7.7)

12.10 g of Tris-hydroxy methyl amino methane was dissolved in 80.00 ml of in HCl in addition to made up to 1 Litter by distilled water. The pH is adjusted in the direction of 7.7.

(ii). 0.02M Adenosine Tri-phosphate Solution

11.023 mg of free ATP was dissolved into 1.00 ml of distilled water (w/v). This solution was arranged new at the time of make use.

(iii). Ammonium Molybdate Solution

25.00 g of ammonium molybdate was dissolve in 100.00 ml (w/v) of distilled water. In 1 L flask, 300.00 ml of 10 N Sulphuric acid (83.10 ml of concerted sulphuric acid of 36 N was made up to 300.00 ml with distilled water) was in use as well as to this, ammonium molybdate was additional plus made up and about to 1 Litter by means of distill water.

(iv). 5% Tri-chloro Acetic Acid Solution

5.0 g of TCA is dissolve in 100.00 ml of distilled water (w/v).

(v). 15% Sodium Bi-sulphite Solution

30.00 g of sodium bi-sulphite is dissolve in 200.00 ml of distilled water (w/v). If the resolution is turbid, it is allowable to position for a number of days also filtered.

(vi). 20% Sodium Sulphite Solution

20.00 g of anhydrous sodium sulphite was dissolve in 100.00 ml of distill water (w/v).

(vii). ANSA Reagent (1-amino-2-naphthol-4-sulphonic acid)

0.50 g of ANSA was dissolve in 195 .00 ml of 15% Sodium bi-sulphite solution into a brown container. Toward this, 5.00 ml of 20% Sodium sulphite solution was added furthermore mixed fine. This was ready freshly used for each use.

(viii). 100 mM Sodium Chloride Solution

584.00 mg of NaCl was dissolve inside 100.00 ml of distill water (w/v).

(ix). 100 mM Potassium Chloride Solution

746.00 mg of KCl was dissolve inside 100.00 ml of distill water (w/v).

Standard preparation

0.315 g of pure monopotassium phosphate was dissolve into distill water moreover transfer quantitatively to 1 Litter volumetric flask. Subsequently 10.00 ml of 10 N sulphuric acids was supplementary to this as well as dilute to the mark by resources of distilled water in addition toward varied fine. These solutions contain 0.40 mg of phosphorus into 5.00 ml.

Procedure used for standard graph preparation

From the standard solution, 0.10 ml, 0.20 ml, 0.30 ml, 0.40 ml, 0.50 ml, 0.60 ml, 0.70 ml, 0.80 ml, 0.90 ml, as well as 1.00 ml of the solution be taken, in that order in 10 test tubes, which correspond just before their individual concentration of 8, 16, 24, 32, 40, 48, 56, 64, 72 along with 80 /µg of phosphorus. Toward each test tube, distill water was additional in addition to made positive to 1.00 ml. subsequently 1.00 ml of ammonium molybdate furthermore 0.40 ml of ANSA be added to all tube in addition to the blue colour - formed be immediately read next to blank at 680 nm use Spectronic – 20D.

In support of scheming the normal graph, the OD values be taken on the Y axis in addition to the attention of phosphorus all along the X axis. Dissimilar concentrations moreover their respective OD values were plotted and points obtained were joined by a straight line.

Procedure used for enzyme assay

Clean test tubes are taken as well because labelled 'Blank' plus 'Test'. To each tube, 0.30 ml of Tris-HCl buffer (PH 7.5), 0.10 ml of 0.02 M ATP, 0.10 ml of 100 mM NaCl and 0.10 ml of KCl (For Na^+, K^+ also-ATPase) be supplementary.

In that case, 0.10 ml of distill water was further to blank in addition to to the test 0.10 ml of tissue extract/ plasma was added to the particular tubes. The response mix was varied as well as incubates in water bath at 37°C for 15 proceedings. Once the incubation period, the response was completed with 2.0 ml of 5% TCA. The tubes are kept at 4°C for 30 minutes as well as centrifuged used for 5 minutes by 500 rpm.

Toward the supernatant, 1.00 ml of Ammonium molybdate into adding to 0.40 ml of ANSA reagent were additional in adding to allowed to position for 10 minutes, at room temperature. The strength of the blue colour residential was read by 680 nm against a reagent blank use spectrophotometer (Spectronic 20D, Milton Roy). Apposite standards are as well run through all batch of assay. The enzyme movement was expressed inside terms of micrograms/ micro liters of inorganic phosphorus produced per hour, per gram of tissue/ per litter of plasma.

Calculation

Step 1.

The OD value of both Controls along with experimental samples was patent on the Y axis of the normal graph as well as it was extrapolated to the matching phosphorus absorption going on the X axis.

Step 2.

The Na^+, K^+, ATPase enzyme activity was calculated as for every the subsequent formula.

For gills

$$\text{Activity} = \text{Concentration of phosphorus} \times \frac{\text{Total volume taken for homogenisation}}{\text{Volume taken}} \times 4 \times \frac{1000}{\text{Weight of}} \times \frac{1}{1000} \text{ expressed in } \mu g/h/g \text{ tissue.}$$

for analysis tissue taken

For plasma

$$\text{Activity} = \text{Concentration of Phosphorus} \times \text{Volume taken for analysis} \times 4 \times \frac{1}{1000}$$ expressed in µl /h/L of plasma.

Estimation of Acid Phosphatase

Acid phosphatase activity is projected in the method of Moctified King's method of King in addition to Jagatheesan (1959) by means of Diagnostic Reagent Kit supplied by Stangen Immuno diagnostics, Hyderabad and India.

Reagents

i. Substrate
ii. Buffer solution pH 5.0
iii. Colour reagent
iv. Phenol standard 10 KA units

Procedure

Reagent preparation, single vial of substrate (i). Be reconstituted among 5.5 ml of buffer solution in addition to (ii).Mixed well. Four test tubes be taken as well as named since Blank (B), Standard (S), Control (C), along with Test (T). Toward each tube 1.0 ml of buffered substrate was supplementary. Then 3.2 ml, 3.1 ml, 3.0 ml also 3.0 ml of deionized water was further in the direction of test tubes Blank (B), Standard (S), Control (C), moreover Test (T) correspondingly. They are varied very well moreover incubated at 37°C for 3 minutes. Subsequently 0.2 ml of plasma be added to the test (T) tube in addition to 0.1 ml of phenol standard was added to the tube standard (S). The more than solutions are mixed fine in addition to incubated at 37°C for 60 minutes. Then 2.0 ml of plasma was additional to control tube (C). The more than reagents be mix well in addition to the absorbance of Blank (B), Standard (S), Control (C) and Test (T) next to distill water were calculated into spectrophotometer by 510 nm.

Calculation

$$\text{ACP activity in KA Units} = \frac{\text{A of (T) - A of (C)}}{\text{A of (S) - A of (B)}} \times 5$$

Estimation of Alkaline Phosphatase

Alkaline phosphatase movement was predictable in Kind and King's (1954) Method by means of Diagnostic Reagent Kit complete by means of Stangen immunodiagnostics, Hyderabad, and India.

Reagents

i. Buffered substrate,
ii. Colour reagent,
iii. Phenol standard (10 KA units).

Reagent preparation

Buffer substrate be reconstituted with 4.5 ml of distill water as well as mixed well.

Procedure

Four test tubes *viz.,* Blank (B), Standard (S), Control (C) in addition to Test (T) were in use moreover to each test tube 1.0 ml of functioning buffered substrate be supplementary. Then 3.1 ml deionized water was additional to the Blank (B), moreover 3.0 ml of deionised water was added in the direction of the rest of three tubes. The samples are incubated for 3 minutes, by 37° C. In that case 0.1 ml of (Plasma) was added to Test (T) along with 0.1 ml of Phenol Standard (S) was added to the Standard (S). Then 0.1 ml of plasma was added to the Control (C) tube. Once more, the samples are incubated for 15 minutes at 37°C in addition to 2.0 ml of colour reagent was added to all of four tubes. The higher than samples be mixed fine after each adding of reagent as well as absorbance be calculated for Blank (B), Standard (S), Control (C) also Test (T) next to deionised water by means of spectrophotometer at 510 nm.

Calculation

$$\text{ALP in KA units} = \frac{\text{A of (T)} - \text{A of (C)}}{\text{A of (S)} - \text{A of (B)}} \times 10$$

Histopathological studies

To study the amount of cellular damage caused in the aluminium within gill, tissue of Fish *Cyprinus carpio var. communis* control also experimental groups be present fixed into Bouin's fluid used for 24 hours. Later than 24 hours, the normal histological method was follow with the method of Gurr, (1959). The tissues are dehydrated during a grade series of alcohol, clean within xylol also embedded into paraffin wax (58°-60°C). Using a revolving microtome, 6μ thick sections are cut. The section is present deparaffinized within xylol, passed during descending grade of alcoholic cycle also next washed by distilled water. Subsequently the sections are stained by Heidenhain's iron haematoxylin also counterstained by aqueous eosin. Stain sections be mounted into DPX in support of microscopic observations.

RESULTS

Exposure of Aluminium on the studies of *Cyprinus carpio var. communis*

The fresh water Fish, *Cyprinus carpio var. communis* shows the abnormal pattern when the Fish exposed to sub lethal concentration of Aluminim. The changes

are manifested by eradicating swimming, restlessness, rapid jerky movement, decreased feeding rate and difficulty in respiration, increased fin movement. The Fish came to the surface of water again and again to engulf the atmospheric air. During the early period of exposure, the Fish try many times to escape from the trough of the toxic aquatic medium.

Change in Electrolyte in addition to acid-base balance into the blood moreover plasma of fish, *Cyprinus carpio var. communis* showing to acute aluminium stress be accessible in (Table 2 also Fig. 1). Blood sodium and plasma chloride level in the experimental fish showed (-12.64%) as well as (-57.01%) decrease from to facilitate of their control groups, correspondingly. Plasma chloride into investigational group showed near fivefold reduces than that of sodium. However, Potassium level greater than before by (+10.82%) than so as to of the control. Also both pH in addition to Bi-carbonate level of blood of fish decline in the investigational group compare to their personal control generous a value of (-0.93%) in addition to (-46.15 %) correspondingly (Table 3 also Fig. 2). All the values are found to be important by 5% level.

Sub-lethal management of fish among aluminium sulphate cause an original decline inside blood sodium level beginning its control give a decrease of (-22.94%) at the ending of 7th days (Table 4 and Fig. 3). However, after 7th days, blood sodium level greater than before steadily showing (+22.57%) into the tested group over to of its control on the finish of 35th days. Another blood sodium level values for 14th, days value (+2.63%), 21st days value (+11.24%), and 28th days value (+18.76%).

Table 2. Sodium, Potassium in addition to Chloride Levels into The Blood as well as Plasma of Cyprinus *Carpio Var. Communis* Exposed in the direction of Acute Aluminium Sulphate Toxicity.

S. No	**Parameters**	**Control**	**Experimental**	**Percent Change**	**Calculated 't'**
01.	Sodium (mmol/L)	87.000 ± 1.581	76.000 ± 1.140	-12.64	5.6429*

02.	Potassium (mmol/L)	7.850 ± 0.074	8.700 ± 0.352	+10.82	2.3627*
03.	Chloride (mmol/L)	114.000 ± 1.304	49.000 ± 1.892	-57.01	28.2688*

Values are mean ± S. E. of five individual observations,
- Denotes percent decrease over control,
+ Denotes percent increase over control,
*Values are significant at 5% level,
Degrees of freedom at 8 to 0.05 = 2.306.

Fig.1. Sodium, Potassium in addition to Chloride levels in the Blood as well as Plasma of *Cyprinus carpio var. communis,* exposed in the direction of Acute Aluminium Sulphate toxicity.

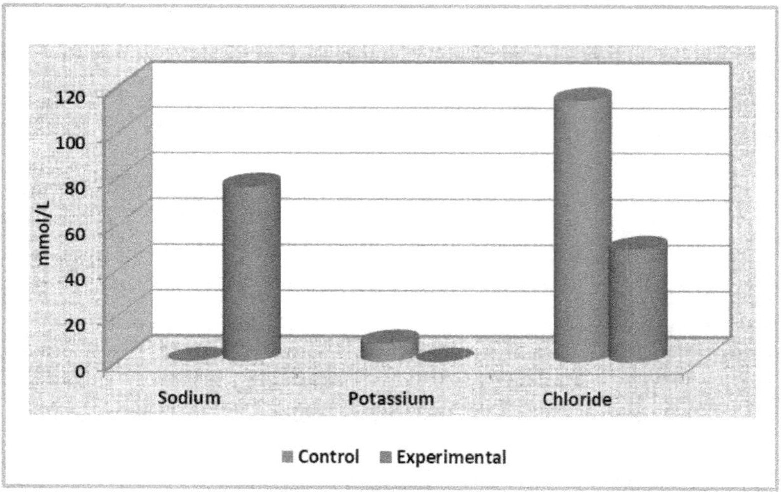

Table 3. pH in addition to Bi-carbonate levels into the blood of *Cyprinus carpio var. communis,* exposed in the direction of Acute Aluminium Sulphate toxicity.

S.No	Parameters	Control	Experimental	Percent change	Calculated 't'
01.	pH (Units)	7.010 ± 0.005	6.945 ± 0.017	-0.93	3.7696*

02.	Bi-carbonate (mmol/L)	3.900 ± 0.212	2.100 ± 0.158	-46.15	6.8034*

Values are mean ± S. E. of five individual observations,
-Denotes percent decrease over control,
* Values are important at 5% level,
Degrees of freedom at 8 to 0.05 = 2.306.

Fig. 2. pH in addition to Bi-carbonate levels into the blood of *Cyprinus carpio var. communis,* exposed in the direction of Acute Aluminium sulphate toxicity.

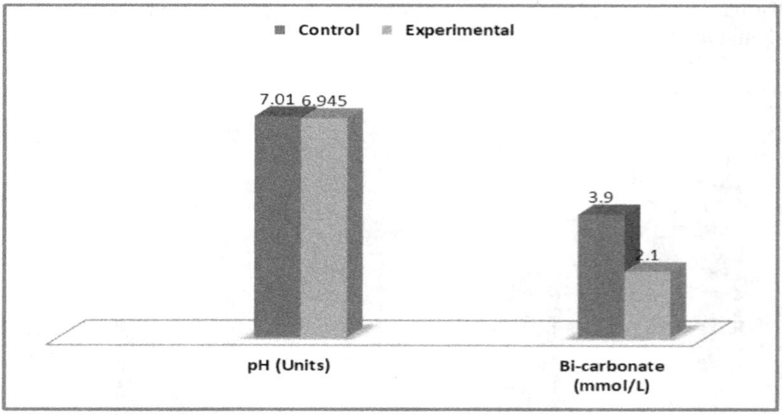

Table 4. Sodium content into the blood of *Cyprinus carpio var. communis,* exposed in the direction of sub-lethal concentration of Aluminium sulphate used for varying periods.

S.No	Exposure period (In Days)	Blood Sodium level in mmol/L		Percent Change
		Control	Treatment	
01.	07	87.200 ± 1.855	67.200 ± 1.050	-22.94
02.	14	76.000 ± 1.226	78.000 ± 1.582	+2.63
03.	21	89.000 ± 1.644	99.000 ± 0.895	+11.24
04.	28	84.200 ± 0.736	100.000 ± 2.026	+18.76

05.	35	82.400 ± 2.136	101.000 ± 1.141	+22.57

Values are mean ± S. E. of five individual observations,
'-' Denotes percent decrease over control,
'+' Denotes percent increase over control.

Fig.3. Sodium content into the blood of *Cyprinus carpio var. communis,* exposed in the direction of sub-lethal concentration of Aluminium Sulphate used for varying periods.

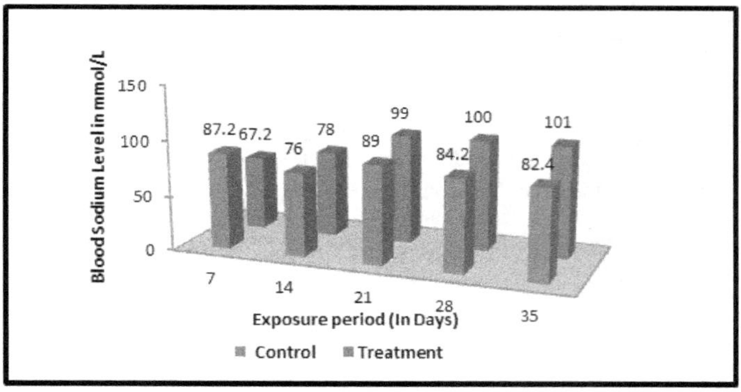

While the periods mean be analysed irrespective of the treatment 7th, 14th and 21st, 28th in addition to 35th days be on par among each other in addition to varied considerably starting every one other. When management means be analysed, the Na^+ level different appreciably than to facilitate of the control. The study clearly depicts to Na^+ level different appreciably than to the control.

Table 5 also Fig. 4; give the data scheduled potassium levels starting the blood of fish showing to the sub-lethal absorption of Aluminium sulphate. The K^+ level improved from (+14.10%) at the finish of 7th days to (+30.39%) at the conclusion of 35th days. While control alone was analysed, 7th in addition to 21st days, 21st also 14th days, 14th also 35th days furthermore 28th also 35th days were on par with each other. The rest of the combination varied significantly between them.

When behavior groups only were analysed 14th also 28th days, 28th also 35th days be present on par with means of every one supplementary. The rest of the combination various significantly starting these groups moreover between themselves.

Table 5. Potassium level into the blood of *Cyprinus carpio var. communis,* exposed in the direction of sub-lethal concentration of Aluminium Sulphate used for varying periods.

S.No	Exposure Period (In Days)	Blood Potassium level in mmol/L		Percent Change
		Control	Treatment	
01.	07	7.800 ± 0.382	8.900 ± 0.246	+14.10
02.	14	9.000 ± 0.162	11.800 ± 0.229	+31.11
03.	21	8.300 ± 0.201	10.300 ± 0.377	+24.10
04.	28	10.600 ± 0.503	13.000 ± 1.026	+22.64
05.	35	10.200 ± 0.597	13.300 ± 0.165	+30.39

Values are mean ± S.E. of five individual observations,

'+' Denotes percent increase over control.

Fig.4. Potassium content into the blood of *Cyprinus carpio var. communis,* exposed in the direction of sub-lethal concentration of Aluminium sulphate used for varying Periods.

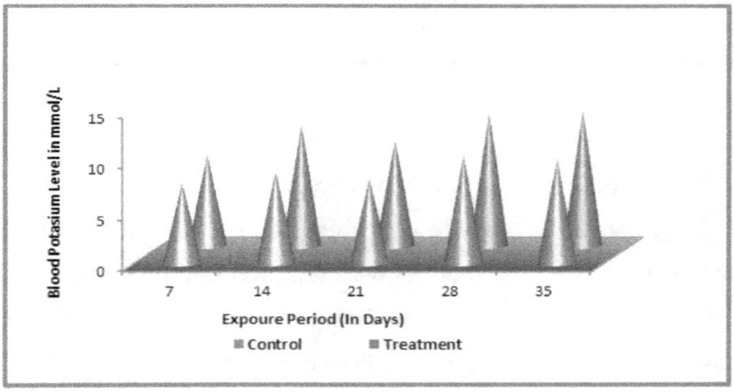

When periods signify was analysed, irrespective of their treatment, every one the contact periods different from each other except 28th in addition to 35th days which be on par through every one other. Once treatment means be analysed blood K^+ level diverse extensively than to facilitate of the control thus representing that K^+ level improved at some point in sub-lethal treatment through aluminium. Chloride satisfied inside the plasma of every one treat fish decline drastically by, (-55.82%) as of control principles after 7th days (Table 6 also Fig. 5). Though it improved slightly give a percent decline of (-44.03%) starting control value at the finish of the 35th days.

Investigation of period's means irrespective of treatment showed to the 7th also 14th days is on par among each other excluding varied starting that of the rest of the contact periods, which too different between themselves. Similarly, study of handling means show that sub-lethal attention of Aluminium sulphate cause a change into Cl^- level which different significantly than to of the control, representative the Aluminium sulphate be extremely toxic in depressing Cl^- level into fish by sub-lethal strain.

Table 6. Chloride level into the plasma of *Cyprinus carpio var. communis,* exposed in the direction of sub-lethal concentration of Aluminium Sulphate used for varying periods.

S.No	Exposure Period (In Days)	Plasma Chloride level in mmol/L		Percent change
		Control	Experimental	
01.	07	112.600 ± 0.759	49.750 ± 0.576	-55.82
02.	14	106.400 ± 1.611	56.000 ± 1.845	-47.57
03.	21	104.800 ± 1.752	71.310 ± 0.457	-31.82
04.	28	126.400 ± 1.072	72.180 ± 0.295	-42.90
05.	35	134.000 ± 1.492	75.000 ± 1.674	-44.03

Values are mean ± S. E. of five individual observations,
'-' Denotes percent decrease over control.

Fig.5. Chloride content into the blood of *Cyprinus carpio var. communis,* exposed in the direction of sub-lethal concentration of Aluminium sulphate used for varying periods.

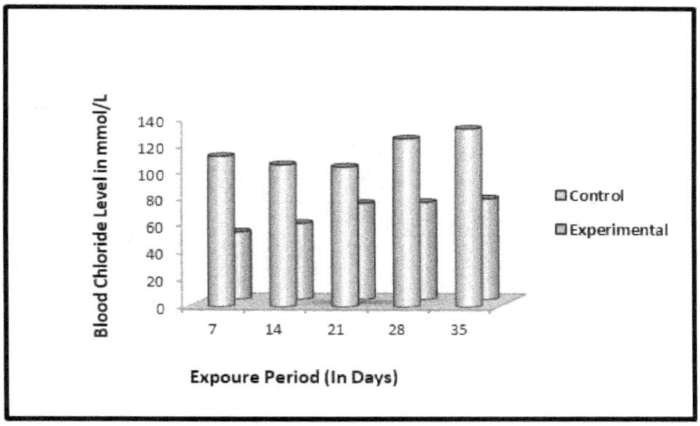

Blood pH stage of *Cyprinus carpio var. communis* bigger up to the finish of 21st days at some point into sub-lethal Aluminium treatment (+2.48%) with relative in the direction of the control (Table 7 and Fig. 6). After 21st days, it however, declined give a reduction of (-3.07%) than the reference value by the finish of 28th days. Another time, pH level improved slightly on the finish of 35th days exhibiting a percent refuse of (-0.95%) starting control value.

When control grouping single-handedly was tested 7th also 14th days, 14th also 21st days, 21st, 28th plus 35th days were going on par by means of every one other, the rest of arrangement organism significantly dissimilar from every one other. Into the case of sub-lethal behavior of aluminium, 7th, 14th, 28th also 35th days treatment were on par by means of every one other excluding, 7th in addition to 14th, 28th days various considerably from so as to of 21st days. Going on the additional give, 35th days was going on par in 21st days.

Investigation of period's means, irrespective of the treatment exposed to 7th, 14th in addition to 28th days, 14th, 28th as well as 35th days also 21st, 28th plus 35th days were on par by means of each one other. The rest of the combination being considerably dissimilar from every one other. Investigations of treatment irrespective of the periods, blood pH do not differ considerably from to the control. Therefore analysis shows to pH level into the investigational group do not differ from the control all through sub-lethal treatment.

Table 7. Blood pH of *Cyprinus carpio var. communis,* exposed in the direction of sub-lethal concentration of Aluminium sulphate.

S.No	Exposure Period (In Days)	Blood pH in Units		Percent Change
		Control	Experimental	
01.	07	7.309 ± 0.007	7.409 ± 0.010	+1.37
02.	14	7.400 ± 0.036	7.486 ± 0.039	+1.18
03.	21	7.541 ± 0.035	7.728 ± 0.062	+2.48
04.	28	7.614 ± 0.109	7.380 ± 0.065	-3.07
05.	35	7.614 ± 0.113	7.452 ± 0.104	-0.95

Values are mean ± S. E. of five individual observations,
'-' Denotes percent decrease over control,
'+' Denotes percent increase over control.

Fig.6. Blood pH of *Cyprinus carpio* var. *communis,* exposed in the direction of sub-lethal concentration of Aluminium Sulphate used for varying periods.

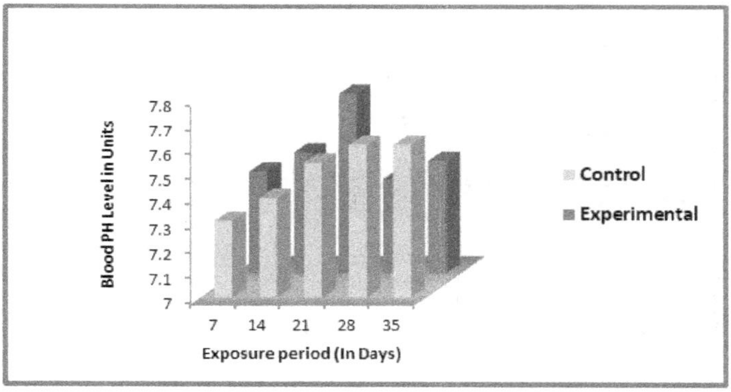

Blood HCO_3^- level increased initially giving (+30.45%) also (+30.00%) after 7th plus 14th days treatment, respectively (Table -8 also Fig.7) which then declined since the exposure period was extended give a percent decline of (-40.54%) by the finish of 35th days.

Table 8. Bi-carbonate level into the blood of *Cyprinus carpio var. communis,* exposed in the direction of sub-lethal concentration of Aluminium sulphate used for varying periods.

S.No	Exposure Period (In Days)	Blood Bi-carbonate level in mmol/L		Percent Change
		Control	Experimental	
01.	07	2.300 ± 0.271	3.000 ± 0.963	+30.45
02.	14	3.000 ± 0.708	3.900 ± 0.286	+30.00
03.	21	3.020 ± 0.328	2.900 ± 0.516	-3.97
04.	28	3.000 ± 0.791	1.800 ± 0.258	-40.00

05.	35	3.700 ± 0.488	2.200 ± 0.443	-40.54

Values are mean ± S. E. of five individual observations,
'-' Denotes percent decrease over control,
'+' Denotes percent increase over control.

Fig.7. Bi-carbonate levels into the blood of *Cyprinus carpio var. communis,* exposed in the direction of sub-lethal concentration of Aluminium Sulphate used for varying periods.

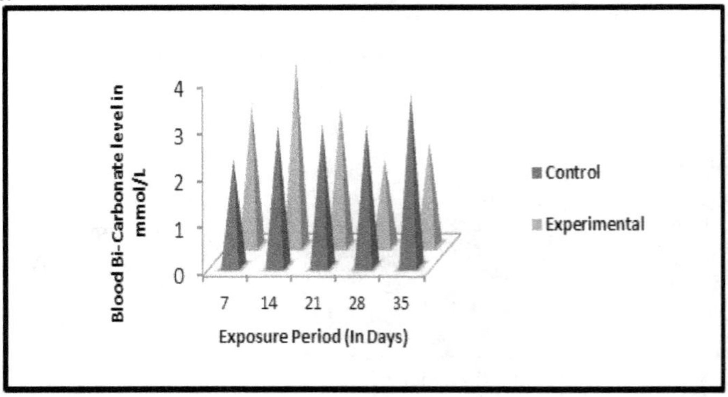

Change into haematological profiles of fish, *Cyprinus carpio var. communis,* in acute aluminium exposure be given inside (Table 9 also Fig. 8), Haemoglobin satisfied plus haematocrit value decreased since the control level in (-11.41%) in addition to (-7.57%), correspondingly. Still their decrease was not important. Erythrocyte also leucocytes counts be altered significantly starting the control charge in register (-15.40%) decrease also (+25.01%) increase, correspondingly.

Erythrocyte index since MCHC declined into the investigational groups as compared toward to of control give a percent reduce of (-3.87%), as MCH also MCV improved compared in the direction of control in (+5.74%) also (+9.62%), in that order. However, statistical study revealed to these changes was not important.

Table 10 also Fig. 9. give the data resting on changes into haemoglobin contented of the fish blood at some stage in sub-lethal aluminium management. At the finish of 7th days a marked decline inside haemoglobin contented (-57.16%) was observed into the investigational fish which progressively recovered throughout the succeeding exposure period presentation a reduce of (-43.20%), (-34.77%), (-29.59%) also (-21.90%) after 14th, 21st, 28th also 35th days, respectively.

Table 9. Changes into the haematological profiles of *Cyprinus carpio var. communis,* exposed in the direction of acute Aluminium sulphate toxicity.

S.No	Parameters	Control	Experimental	Percent Change	Calcula ted 't'
01.	Haemoglobin (g/dl)	3.952 ± 0.1434	3.502 ± 0.1583	-11.41	2.1098
02.	Haematocrit (%)	5.302 ± 0.2552	4.902 ± 0.2917	-7.57	1.0330
03.	Erythrocytes (Million / cubic mm)	1.302 ± 0.0243	1.102 ± 0.0608	-15.40	3.0681*
04.	Leucocytes (Thousands /cubic mm	20.152 ± 91.7015	25.182 ±574.9261	+25.01	5.1472*
05.	MCHC (g / dl)	75.442 ± 5.6628	72.536 ± 5.576	-3.87	0.3659
06.	MCH (Picograms)	30.410 ± 1.1453	32.150 ± 2.0367	+5.74	0.7449
07.	MCV (Cubic micra)	40.742 ± 1.4391	44.652 ± 1.6544	+ 9.62	1.7845

Values are means ± S.E. of five individual observations,
- Denotes percent decrease over control,
+ Denotes percent increase over control, Values are significant at 5% level,
Degrees of freedom by 8 to 0.05 = 2.306.

Fig.8. Changes into the Haematological profiles of *Cyprinus carpio var. communis* exposed in the direction of 24 hours LC_{50} concentration of Aluminium sulphate.

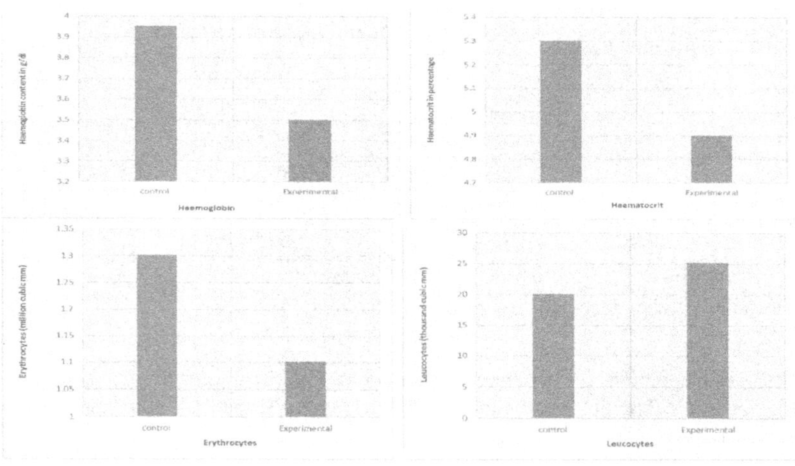

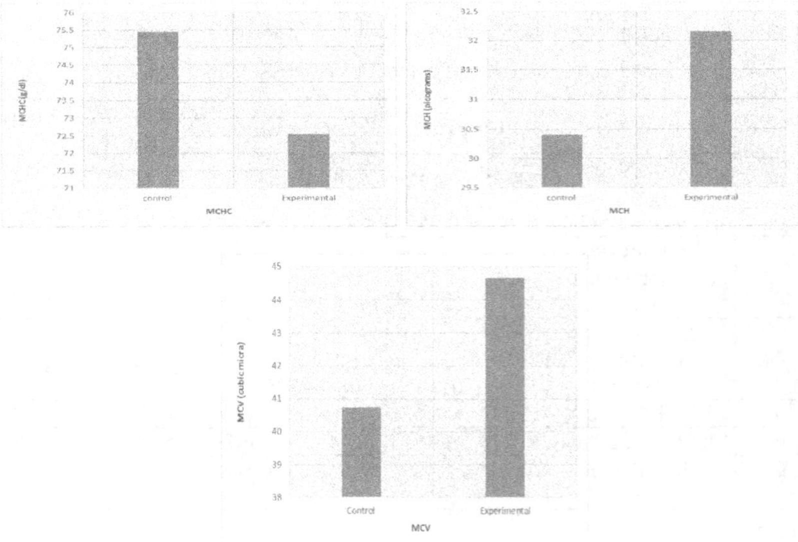

Table 10. Haemoglobin content into the blood of *Cyprinus carpio var. communis,* exposed in the direction of sub-lethal concentration of Aluminium sulphate used for varying periods.

S.No	Exposure Period (In Days)	Blood Haemoglobin Content in g/ dl		Percent Change
		Control	Treatment	
01.	07	2.802 ± 0.166	1.202 ± 0.102	-57.16
02.	14	2.642 ± 0.602	1.502 ± 0.034	-43.20
03.	21	2.822 ± 0.134	1.842 ± 0.083	-34.77
04.	28	2.302 ± 0.132	1.622 ± 0.039	-29.59
05.	35	2.562 ± 0.053	2.002 ± 0.034	-21.90

Values are means ± S.E. of five individual observations

'-' Denotes percent decrease over control

Fig.9. Haemoglobin content into the blood of C*yprinus carpio* var. *communis,* exposed in the direction of sub-lethal concentration of Aluminium Sulphate used for varying periods.

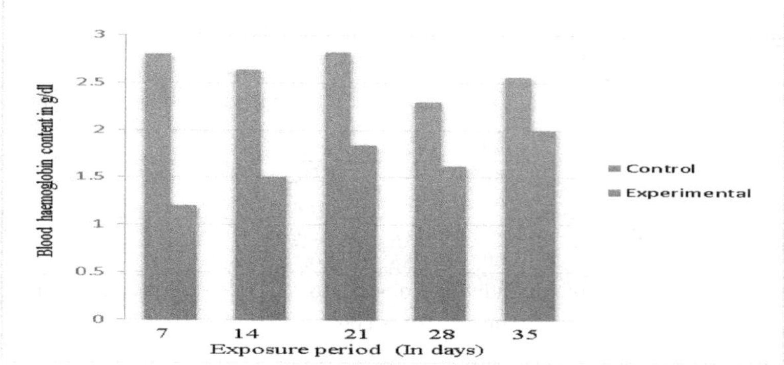

While the control group only was tested 7th plus 28th, 14th also 28th, 21st plus 28th, 7th also 35th days were considerably dissimilar whereas 7th and 14th, 14th plus 21st, 14th also 35th, 28th in addition 35th days be on par by means of every one other. When management grouping alone was tested, 7th days varied considerably from all additional contact periods, while 14th in addition to 28th days, 21st plus 28th days and 21st also 35th days were on par by means of each other in addition to them different from the break of the combination.

As soon as period means be analysed, 7th, 14th in addition to 28th days, 21st also 35th days were not considerably dissimilar among every one other excluding they various with the rest of their combination. After treatment means be analysed, together the control in addition to treatment different considerably among every one other, thus presentation to sub-lethal aluminium inform poisonous effect on top of haemoglobin stage intothe fish. Contact of fish in the direction of sub-lethal aluminium sulphate resulted into a (+11.65%) increase during the haematocrit value of fish on the end of 7th days (Table 11 also Fig. 10). Though after 7th days, the haematocrit price declined charitable a percent refuse of (-30.02%), excluding at the finish of the 21st days, the haematocrit reach a value or (-0.002%) present. On end of the 28th days the haematocrit value another time decreased presentation (-17.67%). Going on contrary, by the end of the 35th days, haematocrit worth improved by (+50.02%) compare in the direction of the control.

Table 11. Haematocrit values of *Cyprinus carpio var. communis,* exposed in the direction of sub-lethal Aluminium sulphate toxicity used for varying periods.

S.No	Exposure Period (In Days)	Blood Haematocrit (%)		Percent change
		Control	Treatment	
01.	07	4.302 ± 0.124	4.802 ± 0.257	+11.65
02.	14	5.002 ± 0.356	3.502 ± 0.356	-30.02
03.	21	4.000 ± 0.356	4.000 ± 0.356	-0.02
04.	28	5.102 ± 0.102	4.202 ± 0.124	-17.67
05.	35	4.002 ± 0.356	6.002 ± 0.160	+50.02

Values are mean ± S. E. of five individual observations, '-' Denotes percent decrease over control, '+' denotes percent increase over control.

Fig.10. Haematocrit values of *Cyprinus carpio var. communis,* exposed in the direction of sub-lethal concentration of Aluminium sulphate used for varying periods.

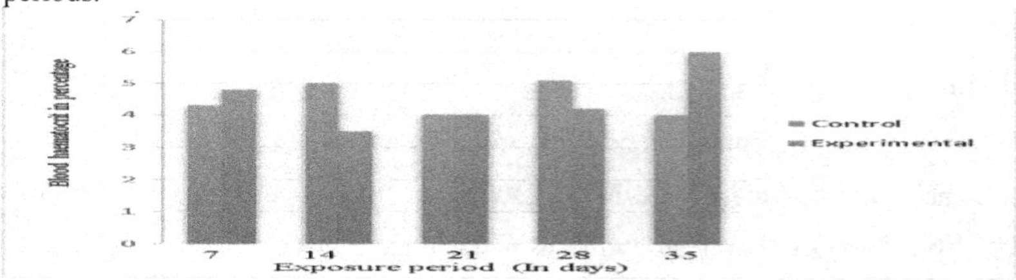

Study of control group unaccompanied for unreliable period's show the 7th, 21st also 35th days, 7th, 14th, in addition to 28th days were on par with each other, while 14th and 21st, 14th and 35th, 21st and 28th, 28th and 35th were significantly different. Investigation of treatment collection alone for varying period's show that 7th and 28th, 14th, 21st and 28th days were on par by each other. On the additional hand, 35th days varied considerably from the rest of the publicity periods.

Study of period means irrespective of the action indicated that 7th, 14th and 21st, 7th, 14th and 28th and 7th, 28th and 35th days be on par with every one other, although the rest of the combination being considerably dissimilar from each one other. Investigation of treatment means irrespective of the contact period exposed that the group's *viz.,* control also test were going on par with every one other.

RBC cells count of the blood sample of *Cyprinus carpio var. communis* after 7th days management was concentrated severely by (-45.71%). However, there was a gradual recovery during sub-sequent exposure periods give a percent refuse of (-38.22%), (-30.78%), (-27.27%) and (-21.56%) after 14th, 21st, 28th and 35th days, in that order (Table 12 and Fig. 11).

Table 12. Erythrocytes count into the blood of *Cyprinus carpio var. communis,* exposed in the direction of sub-lethal concentration of Aluminium sulphate used for varying periods.

S.No	Exposure Period (In Days)	Erythrocytes Count In Millions / Cu.mm		Percent Change
		Control	Treatment	
01.	07	0.593 ± 0.019	0.323 ± 0.015	-45.71
02.	14	0.803 ± 0.027	0.497 ± 0.032	-38.22
03.	21	0.912 ± 0.018	0.632 ± 0.034	-30.78
04.	28	0.724 ± 0.111	0.528 ± 0.028	-27.27
05.	35	0.785 ± 0.034	0.616 ± 0.288	-21.56

Values are mean ± S. E. of five individual observations,
'-' Denotes percent decrease over control.

Fig.11. Erythrocytes Count into the blood of *Cyprinus carpio var. communis,* exposed in the direction of sub-lethal concentration of Aluminium Sulphate used for varying periods.

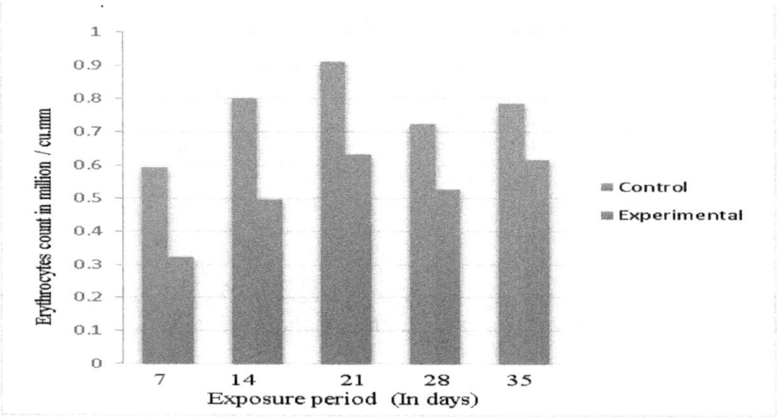

As period means be analysed irrespective of the treatments 14th and 28th, 14th and 35th days be on par through each added while 7th and 21st days different among themselves also varied from so as to of 14th, 28th and 35th days. Study of treatment means show that the investigational groups diverse from the manage depict the aluminium shock on the WBC cell count of the blood. Table 13 also Fig. 12, current the data happen the change in WBC counts into the blood of *Cyprinus carpio var. communis,* expose to sub-lethal attention of aluminium sulphate. The WBC count bigger than to of the control in (+80.99%) on 7th days treatment in addition toward further bigger on 14th and 21st days generous a value of (+133.35% and +130.25%), in to facilitate arrange. immobile the WBC counts refuse gradually keen on the subsequent weeks 28th and 35th days appearance (+83.00% also +50.51%) add to from the control, in that order.

Table 13. Leucocytes count into the blood of *Cyprinus carpio var. communis,* exposed in the direction of sub-lethal concentration of Aluminium sulphate used for varying periods.

S.No	Exposure Period (In Days)	Leucocytes Count In Thousands/ Cu.mm		Percent change
		Control	Treatment	
01.	07	18.502 ± 0.804	33.482 ± 1.319	+80.99
02.	14	19.202 ± 0.666	44.802 ± 1.039	+133.35
03.	21	21.602 ± 1.979	49.732 ± 1.178	+130.25
04.	28	27.902 ± 1.236	51.053 ± 0.808	+83.00
05.	35	19.512 ± 0.656	29.362 ± 0.760	+50.51

Values are mean ± S. E. of five individual observations,
'-' Denotes percent decrease over control,
'+' Denotes percent increase over control.

Fig.12. Leucocytes count in the blood of *Cyprinus carpio var. communis,* exposed to sub-lethal concentration of Aluminium sulphate for varying periods.

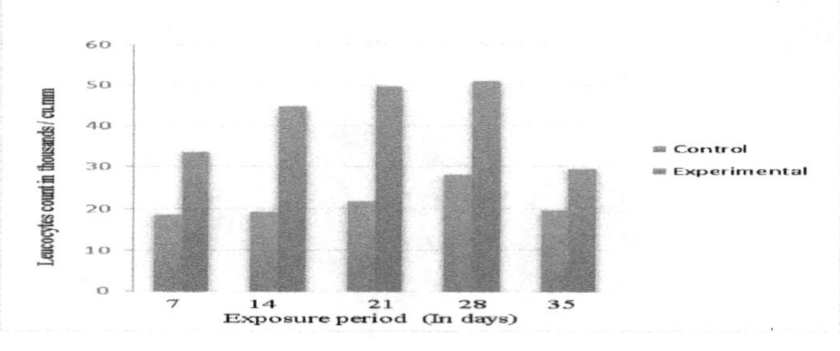

Once period means be analysed 7th and 35th days be on par by means of each additional also various significantly beginning the other period means, while, 14th also 21st days and 28th days various considerably in the middle of themselves. Study of treatment way irrespective of the experience period exposed that the dealing groups dissimilar considerably beginning to facilitate of the control representative sub-lethal pressure of aluminium cause change in the WBC count of fish blood.

The changes occur due to the sub-lethal aluminium exposure in the MCHC value is given in (Table 14 also Fig. 13), MCHC value declined in means of (–61.01%) following 7th days contact also in the sub-sequent weeks the percent refuse was summary giving a importance of (-16.40%), (-34.56%) also (-14.39%) on 14th, 21st and 28th days, into to order. However, subsequent toward the 35th days the percent refuse of the MCHC significance once additional augmented presentation (-49.43%) declined to relative to the control.

When control groups be real analysed use for variable periods 7th, 14th and 35th, 14th and 28th, 21st and 35th days are on par by each other. The rest of the combination various considerably in the middle of each additional. While action group alone be analysed, 7th, 28th and 35th days, 14th and 28th days and 14th and 21st days be on par amongst each other. The 7th and 14th, 7th and 21st, 14th and 35th and 21st and 35th, 21st and 28th days mixed considerably amongst every supplementary.

Table 14.MCHC values of *Cyprinus carpio var. communis,* exposed in the direction of sub-lethal concentration of Aluminium sulphate toxicity used for varying periods.

S.No	Exposure Period (In Days)	Mean Cellular Haemoglobin Concentration In g/dl		Percent change
		Control	Treatment	
01.	07	66.058 ± 3.0484	25.380± 2.6310	-61.01
02.	14	53.928 ± 4.163	45.096 ± 5.658	-16.40
03.	21	73.526 ± 8.868	48.132 ± 6.303	-34.56

04.	28	45.129 ± 2.588	38.669 ± 2.652	-14.39
05.	35	65.992 ± 5.810	33.385 ± 0.817	-49.43

Values are mean ± S. E. of five individual observations,
'-' Denotes percent decrease over control.

Fig.13. MCHC values of *Cyprinus carpio var. communis,* exposed to sub-lethal concentration of Aluminium sulphate used for varying periods.

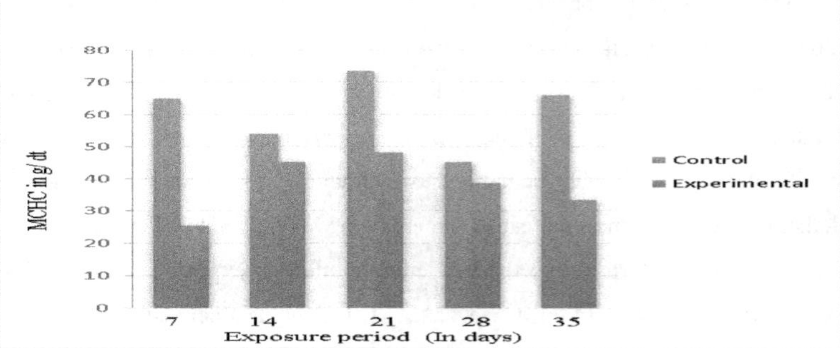

Table 15 also Fig.14. Illustrate the changes going on in the MCH importance due to aluminium pressure in fish blood. The MCH value decrease than to of the control behind 7th days recording (-21.67%). In the subsequent periods, MCH value decrease was very low register (-7.30%), (-2.20%) also (-2.56%), on the 14th, 21st in addition to 28th days, in that order. Whereas on 35th days, MCH value increased compare to the control in (+9.66%).

Study of control group on your own depict that 7th days was important from to facilitate of 14th, 21st, 28th and 35th days which be on par by means of each additional. When management groups be analysed 14th, 21st and 28th days, 14th, 28th and 35th days, 7th, 28th and 35th days be on par with every one added. While 7th and 14th days, 7th and 21st days, 21st and 35th days different considerably between themselves.

Table 15. MCH value of *Cyprinus carpio var. communis,* exposed in the direction of sub-lethal concentration of Aluminium sulphate toxicity used for varying periods.

S. No	Exposure Period (In Days)	Mean Cellular Haemoglobin Value In Picograms		Percent Change
		Control	Treatment	
01.	07	47.574 ± 3.387	37.273 ± 2.375	-21.67

02.	14	33.153 ± 1.668	30.739 ± 1.932	-7.30
03.	21	30.974 ± 1.231	30.298 ± 2.274	-2.20
04.	28	31.911 ± 2.092	31.112 ± 1.725	-2.56
05.	35	32.906 ± 1.459	36.077 ± 1.042	+9.66

Values are mean ± S. E. of five individual observations, '-' Denotes percent decrease over control,'+' denotes percent increase over control,

Fig.14. MCH values of *Cyprinus carpio var. communis,* exposed into the direction of sub-lethal concentration of Aluminium sulphate used for varying periods.

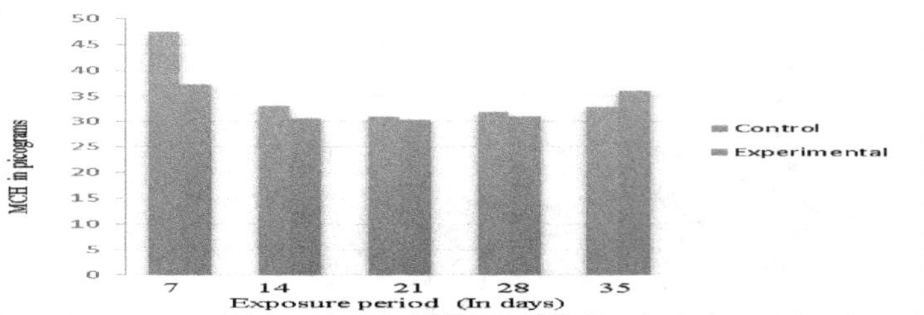

Table 16 and Fig. 15; present the data on change in the action groups by means of stare to MCV value within fish. Present was an evident percent add to of (+105.55%) in the MCV worth in aluminium treat fishes later than 7th days. conversely, the augment in MCV value decrease in the next weeks charitable (+16.11%), (+46.78%) also (+14.11%) on 14th, 21st in addition to 28th days, in that order. Conversely on 35th days, the percent alteration in MCV value another time rose to (+94.73%) relative in the direction of the control.

As the control groups only were experienced for changeable experience periods, 7th, 14th days and 28th days and 21st and 35th days are on parity with every other other than they mixed in the rest of the combination. Also when the behavior groups are analysed 7th and 14th days, 14th and 21st days be on par by each additional, but different considerably starting that of 7th, 28th also 35th days, which as well mixed with themselves.

Table 16. MCV values of *Cyprinus carpio var. communis,* exposed in the direction of sub-lethal concentration of Aluminium sulphate toxicity used for varying periods.

S. No	Exposure Period (In Days)	Mean Cellular Volume Values In Cubic Micra		Percent Change
		Control	Treatment	

01.	07	73.094 ± 3.962	150.224± 9.084	+105.55
02.	14	62.652 ± 4.780	72.732 ± 10.025	+16.11
03.	21	44.278 ± 4.621	64.982 ± 4.641	+46.78
04.	28	70.700 ± 2.275	80.660± 4.566	+14.11
05.	35	50.736 ± 2.800	98.782 ± 6.149	+94.73

Values are mean ± S. E. of five individual observations,

'+' Denotes percent increase over control.

Fig.15. MCv values of *Cyprinus carpio var. communis,* exposed in the direction of sub-lethal concentration of Aluminium sulphate used for varying periods.

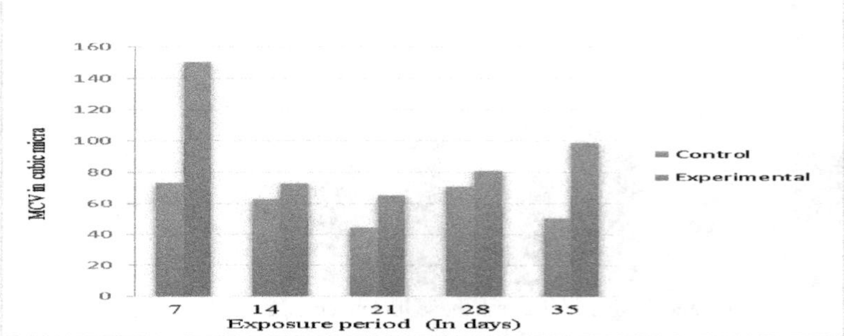

Examination of period way irrespective of the treatment show to 14th and 21st days,14th and 28th and 35th days be on similarity with both other, other than considerably varied beginning the relax of the combination.

While action means be analysed irrespective of the experience period, the management groups different considerably from so as to of the manage collection thus informative to aluminium sulphate be toxic next to sub-lethal stage toward go fish *Cyprinus carpio var. communis.*

Acute absorption of aluminium sulphate cause important reduce into protein level inside the experienced fish than so as to of it's manage presentation a percent reduce of (-21.74) while glucose level greater than before considerably in the investigational fish exhibit a percent boost of (-51.79%), (Table 17, Fig. 16) on the finish of 24 hours' test. Alteration into both the parameters was important on 5 percent stage.

Fish treat through sub-lethal absorption of aluminium sulphate show a decrease of (-28.13%) inside the plasma protein stage beginning to of the manage after 7th days (Table -18 Fig.17). Though, increase, into plasma protein of footage

(+7.14% also +1.05%) is experiential percent, on the finish of 14th in addition to 21st days, in that order. Here the successive weeks, plasma protein show a considerable reduce into their levels inside treat fish compare to of the manage register (-33.33% also -46.82%) refuse after 28th and 35th days, correspondingly.

Table 17. Protein in addition to Glucose levels into the plasma of *Cyprinus carpio var. communis,* exposed in the direction of Acute Aluminium sulphate toxicity.

S. N	Parameters	Control	Experimental	Percent change	Calculated 't'
01	Protein (g/dl)	2.300 ± 0.183	1.800 ± 0.0317	-21.74	2.7116*
02	Glucose (mg/dl)	112.000 ± 1.644	170.000 ± 1.343	+51.79	27.3415*

Values are mean ± S.E. of five individual observations,
'-' Denotes percent decrease over control,
'+' denotes percent increase over control,
Values are significant at 5% level,
Degrees of freedom at 8 to 0.05 = 2.306.

Fig.16. Protein in addition to Glucose levels into the plasma of *Cyprinus carpio var. communis,* exposed in the direction of Acute Aluminium sulphate toxicity.

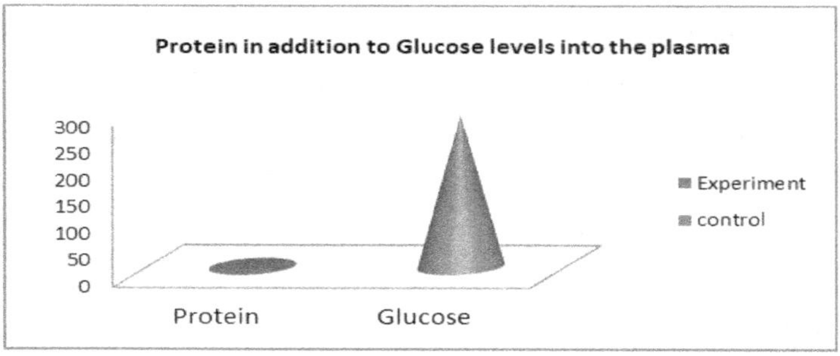

Table 18. Protein level into the plasma of *Cyprinus carpio var. communis,* exposed in the direction of sub-lethal Aluminium sulphate toxicity used for varying periods.

S.No	Exposure Period (In Days)	Plasma Protein Level in g/dl		Percent change
		Control	Treatment	
01.	07	3.200 ± 0.208	2.300 ± 0.502	-28.13
02.	14	4.200 ± 0.188	4.500 ± 0.368	+7.14
03.	21	3.800 ± 0.267	3.840 ± 0.318	+1.05

04.	28	3.600 ± 0.473	2.500 ± 0.280	-33.33
05.	35	4.400 ± 0.767	2.340 ± 0.355	-46.82

Values are mean ± S.E. of five individual observations,
'-' Denotes percent decrease over control,
'+' denotes percent increase over control.

Fig.17. Protein level into the plasma of *Cyprinus carpio* var. *communis,* exposed in the direction of sub-lethal concentration of Aluminium sulphate used for varying periods.

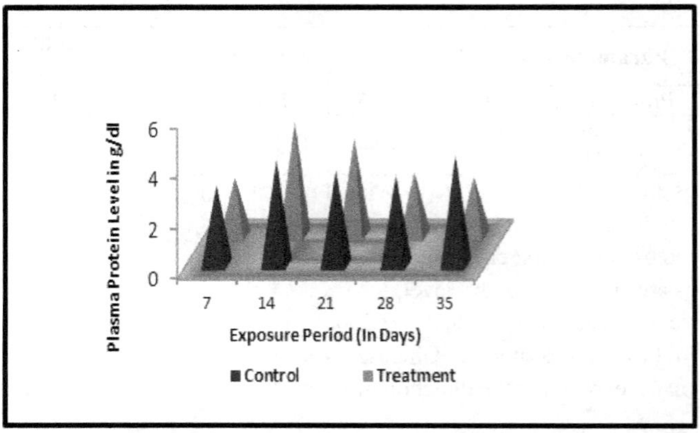

Study of period way irrespective of the behavior show to 7th, 28th and 35th days, 21st, 28th in addition to 35th days in addition to 14th also 21st days be on balance by each additional, as they different considerably from the have a rest of the mixture. Investigation of management resources irrespective of the coverage period show to the behavior mean diverse appreciably beginning so as to of the manage. The over study reveal to facilitate protein level was considerably exaggerated through sub-lethal Aluminium strain.

Table - 19 and Fig.18 provide the data on top of the change within plasma glucose level of fish beginning sub-lethal study. A noticeable rise during the plasma glucose stage was experimental during the treat fish than to facilitate of the manage fish all through the investigational period because fine as it be straight comparative to the period of contact. A preliminary add to of (+39.22 %) was experiential after 7th days, which attempt positive to (113.28 %) following 35th days.

As control group single-handedly was experienced designed for reliable period 14th, 28th days also 7th, 14th, 21st and 28th days be on balance with all other, while the relax of the contact period were considerably dissimilar from persons groups in addition to in the middle of themselves. Also when action alone be analysed used for changeable experience period, only 7th also 14th days be on parity with every one additional, except they different considerably from to of 21st, 28th also 35th days, which also considerably mixed amongst themselves.

Table 19. Glucose level into the plasma of *Cyprinus carpio var. communis,* exposed in the direction of sub-lethal Aluminium sulphate toxicity used for varying periods.

S.No	Exposure Period (In Days)	Plasma Glucose Level in mg/ dl		Percent Change
		Control	Treatment	
01.	07	127.400 ± 7.348	177.500 ± 1.831	+39.22
02.	14	120.000 ± 1.640	180.000 ± 1.722	+50.00
03.	21	125.800 ± 2.220	203.800 ± 1.486	+61.90
04.	28	114.000 ± 1.580	232.500 ± 1.900	+103.95
05.	35	128.000 ± 1.415	273.000 ±1.155	+113.28

Values are mean ± S.E. of five individual observations,
'+' Denotes percent increase over control.

Fig.18. Glucose level into the plasma of *Cyprinus carpio var. communis,* exposed in the direction of sub-lethal concentration of Aluminium sulphate used for varying periods.

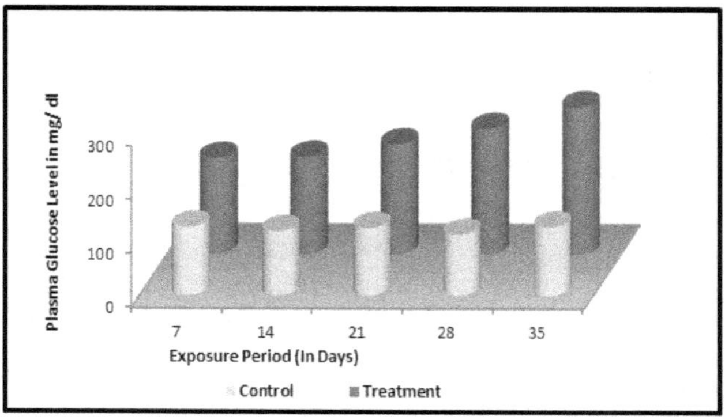

Study of stage way, irrespective of the dealing also exhibits a comparable trend similar to to of the management collection. While action income were analysed irrespective of their publicity period, the behavior group different considerably from the manage group therefore representative to aluminium publicity is poisonous to fish next to sub-lethal levels excessively.

Table 20. also Fig.19. represent the change inside the enzyme behavior of Na^+, K^+, and ATPase during Gills plus plasma also acid in addition to alkaline phosphatase throughout plasma of *Cyprinus carpio var. communis,* presentation to deadly notice of Aluminium sulphate. During gills, the Na^+, K^+, ATPase action decrease considerably presentation (-35.49%) while inside plasma it augmented by (+6.37%) family member in the direction of the control. Mutually acid also alkaline phosphatase movement greater than before during experimental fish, presentation a (+144.66% also +387.69%) respectively.

Table 21. in addition to Fig. 20, give the data under the change into Na^+, K^+, ATPase movement during the gills of fish, *Cyprinus carpio var. communis* showing in the direction of sub-lethal attention of aluminium. The enzyme movement decrease to (-3.54%) following 7th days within treat fish. The seen better days leaning sustained up in the direction of center of the test footage (-18.66%) as well as (-46.92%) decrease on 14th and 21st days, correspondingly. Once 21st days, the enzyme action progressively healthier footage a percent refuse of (-16.02%) and (-8.69%) on the end of 25th and 35th days correspondingly.

Table 20. Na^+, K^+, ATPase activity into the Gills in addition to Plasma also Plasma phosphatases movement of *Cyprinus carpio var. communis,* expose in the direction of Acute Aluminium sulphate toxicity.

S. No	Parameters	Control	Experimental	Percent change	Calculated 't'
01.	Na^+, K^+, ATPase. • Gills (μg/h/gm) • Plasma (μl/h/L)	42.402 ± 1.045 0.0241 ± 0.009	27.362 ± 0.890 0.0256 ± 0.0009	-35.49 + 6.37	10.9817* 1.4769
02.	Acid phosphatase in	22.87 ±	55.902 ± 1.176	+144.66	25.6750*

	KA units	0.529			
03.	Alkaline phosphatase in KA units	14.602 ± 0.178	71.202 ± 1.478	+ 387.69	38.0901*

Values are mean ± S. E. of five individual observations,
 - Denotes percent decrease over control,
+ Denotes percent increase over control,
Values are significant at 5% level,
Degrees of freedom at 8 to 0.05 = 2.306.

Fig. 19. Na^{+}, K^{+}, ATPase activity into Gills and Plasma and Plasma phosphatase activity of *Cyprinus carpio var. communis,* exposed to 24 hours LC_{50} concentration of aluminium sulphate toxicity.

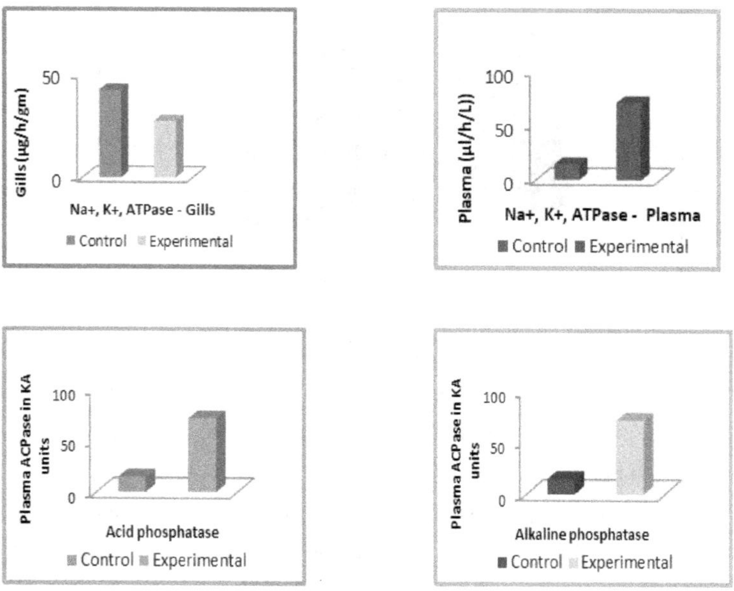

Table 21. Na^{+}, K^{+}, ATPase activity into the Gills of *Cyprinus carpio var. communis,* exposed in the direction of sub-lethal concentration of Aluminium sulphate toxicity used for varying periods.

S.No	Exposure period (In Days)	Gill Na^{+}, K^{+}, ATPase Activity µg/h/g tissue		Percent change
		Control	Treatment	
01.	07	42.002 ± 1.047	40.525 ± 1.260	-3.54
02.	14	36.322 ± 1.230	29.552 ± 0.709	-18.66
03.	21	36.162 ± 1.149	19.202 ± 0.874	-46.92
04.	28	40.002 ± 1.822	33.602 ± 1.056	-16.02
05.	35	42.402 ± 0.714	38.722 ± 1.506	-8.69

Values are mean ± S. E. of five individual observations,
- Denotes percent decrease over control

Fig. 20. Na^{+}, K^{+}, ATPase activity into the Gills of Cyprinus carpio *var.* communis, exposed to sub-lethal concentration of Aluminium sulphate toxicity used for varying periods.

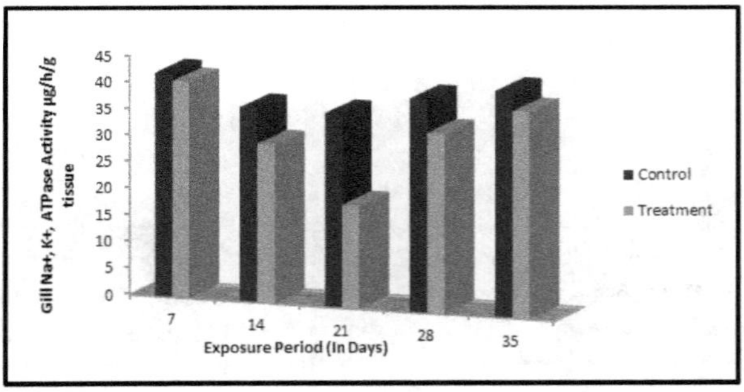

While control group only was analysed used for unreliable experience period 7th, 28th and 35th days be on similarity with all other, excluding they different considerably from to of the 14th and 21st which be too scheduled equivalence by means of both additional.

While behavior group alone be analysed 7th and 35th days be on equivalence by each added but different extensively from with the purpose of 14th, 21st and 28th being which furthermore varied appreciably amongst themselves. A comparable leaning was experiential for episode capital. When behavior means be present analysed irrespective of the disclosure period, together the manage and action differ major by beginning each supplementary representative the harmful achievement of Aluminium within enzyme activity into gills.

Table 22 also Fig.21. Show statistics on Na^+, K^+, ATPase movement in plasma of fish, Cyprinus *carpio var. communis,* expose in the direction of sub-lethal aluminium strain. Na^+, K^+, ATPase activity decrease in (-15.40%) also (-27.09%) after 7th and 14th days, in that order. Though after 21st days enzyme activity into plasma of treated fish all of a sudden better by (+83.12%) than to of the direct. Sub-sequently Na^+, K^+, ATPase action reach to next to regular value of manage soundtrack (-1.30%) reduce also (+6.29%) add to following 28th in addition to 35th days, in that order.

Table 22. Na^+, K^+, ATPase activity into the plasma of *Cyprinus carpio var. communis,* exposed in the direction of sub-lethal concentration of Aluminium sulphate toxicity used for varying periods.

S.No	Exposure Period(In Days)	Plasma Na^+, K^+, ATPase µl/h/L		Percent Change
		Control	Treatment	
01.	07	0.0236 ± 0.0008	0.0200 ± 0.0011	-15.40
02.	14	0.0231 ± 0.0012	0.0169 ± 0.0008	-27.09
03.	21	0.0229 ± 0.0012	0.0417 ± 0.0009	+83.12
04.	28	0.0236 ± 0.009	0.0233 ± 0.0008	-1.30
05.	35	0.0241 ± 0.0012	0.0256 ± 0.0012	+6.29

Values are mean ± S.E. of five individual observations,

- Denotes percent decrease over control,

+ Denotes percent increase over control.

Fig.21. Na^+, K^+, ATPase activity into the plasma of *Cyprinus carpio var. communis,* exposed in the direction of sub-lethal concentration of Aluminium sulphate toxicity used for varying periods.

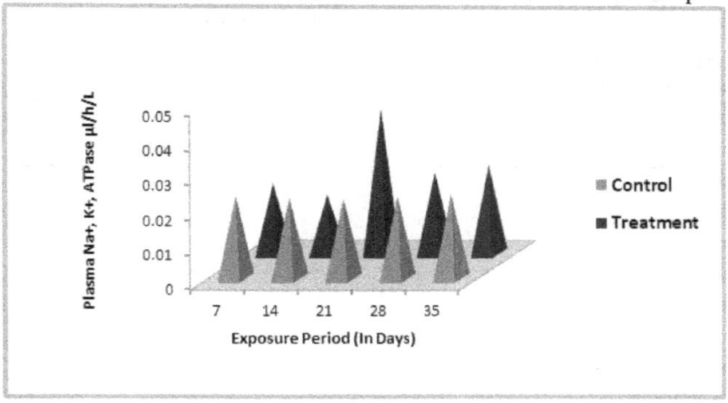

Examination of manage group unaccompanied used for changeable period of contact showed to facilitate the Na^+, K^+, ATPase behavior be on similarity at every one the period. Study of action collection alone exposed only 28th and 35th days is on equivalence with every one other other than varied considerably from to of 7th, 14th and 21st days, which in addition different considerably amongst themselves.

Investigation of stage way, irrespective of behavior resources exposed that 7th in addition to 14th, 7th also 28th plus 28th also 35th days is resting on par between each added, as the relax of the combination various considerably. Study of behavior way irrespective of the contact stage as well show to the behavior group different considerably than the manage.

Change into the plasma acidic phosphatase movement of *Cyprinus carpio var. communis,* exposed to sub-lethal aluminium sulphate be given into (Table 23 also Fig.22.), the enzyme doings bigger to (+96.01%) in the investigational fish following 7th day's management. However, succeeding in the direction of 14th days, it decrease headed for (-19.81%). In the subsequent weeks, another time the enzyme activity considerably better presentation (+79.12%), (+222.11%) also (+285.53%) after 21st, 28th and 35th days, in that order.

while manage group by yourself be tested 7th in addition to 14th, 14th as well as 21st, 28th in addition to 35th days be on balance with every one additional as 7th, 21st in addition to 28th days, 7th, 21st also 35th days various considerably in the middle of themselves. Investigation of management group unaccompanied for changeable period of experience exposed that every one the contact periods mixed considerably among themselves.

Table 23. Acid phosphatase activity into the plasma of *Cyprinus carpio var. communis,* exposed in the direction of sub-lethal concentration of aluminium sulphate used for varying periods.

S. No	Exposure Period (In Days)	Plasma acid phosphatase in KA units		Percent change
		Control	Treatment	
01.	07	24.952 ± 0.592	48.902 ± 1.066	+96.01
02.	14	27.267 ± 0.377	21.862 ± 0.827	-19.81

03.	21	28.592 ± 1.207	51.212 ± 0.841	+79.12
04.	28	21.402 ± 0.746	69.112 ± 0.639	+222.11
05.	35	21.402 ± 0.452	82.502 ± 0.953	+285.53

Values are means ± S. E. of five individual observations,

- Denotes percent decrease over control,

+ Denotes percent increase over control.

Fig.22. Acid phosphatase activity into the plasma of *Cyprinus carpio var. communis,* exposed to sub-lethal concentration of Aluminium sulphate toxicity used for varying periods.

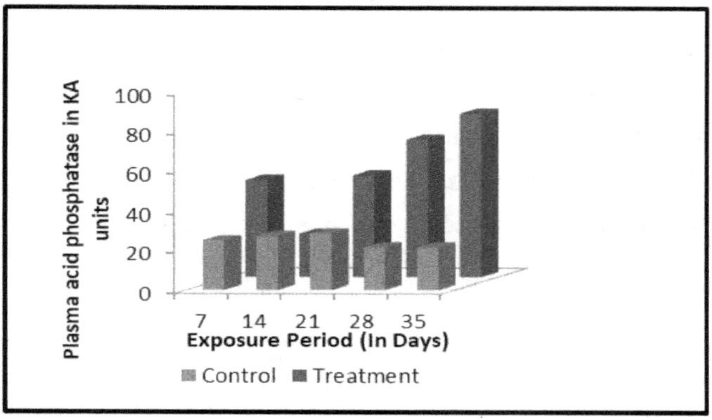

Study of period resources irrespective of the treatment too depict a comparable tendency like to of the management group. As behavior means be analysed irrespective of the disclosure stage, together the manage in addition to behavior groups mixed considerably among themselves.

Table 24. also Fig.23. Demonstrate the change occur in the alkaline phosphatase stage into the plasma of cast a line by the experience of sub-lethal absorption of aluminium sulphate. Alkaline phosphatase action decrease to (-54.35%) on the finish of 7th days, following among the movement steadily improved presentation (-27.72%) on the finish of 14th days after that the worth reach to next to usual by register (+4.63%) add to following 21st days. after that the enzyme commotion decrease presentation (-51.79%) in addition to (-29.91%) after 28th and 35th days, in that order.

After control group alone be experienced, 7th in addition to 21st days, 14th, 28th also 35th days, 21st plus 35th days be on balance with all other though the relax of the combination diverse extensively. As soon as action groups by yourself were analysed 14th in addition to 35th days were on parity through each additional but diverse considerably from to of 7th, 21st in addition to 28th days, which too varied considerably in the middle of themselves.

After period resources be analysed irrespective of the management,7th, 14th in addition to 35th days also 14th and 28th be on similarity with both other, as 7th, 21st in addition to 28th also 35th days various considerably in the middle of every additional. When action resources were analysed irrespective of the experience periods, individually of them various extensively in the middle of all other thus indicating to aluminium impart toxicity resting on alkaline phosphatase movement into the plasma of fish *Cyprinus carpio var. communis.*

Table 24. Alkaline phosphatase activity into the plasma of *Cyprinus carpio var. communis,* exposed in the direction of sub-lethal concentration of Aluminium sulphate toxicity used for varying periods.

S. No	Exposure Period (In Days)	Plasma Alkaline Phosphatase Activity in KA Units		Percent Change
		Control	Treatment	
01.	07	16.602 ± 0.700	7.582 ± 0.310	-54.35
02.	14	14.802 ± 0.855	10.702 ± 0.493	-27.72
03.	21	15.202 ± 0.457	15.902 ± 0.422	+4.63
04.	28	14.102 ± 0.489	13.672 ± 0.832	-51.79
05.	35	14.702 ± 0.304	10.308 ± 0.806	-29.91

Values are means ± S. E. of five individual observations

,- Denotes percent decrease over control,

+ Denotes percent increase over control.

Fig.23. Alkaline phosphatase activity into the plasma of Cyprinus *carpio var. communis,* exposed in the direction of sub-lethal concentration of Aluminium sulphate toxicity used for varying periods.

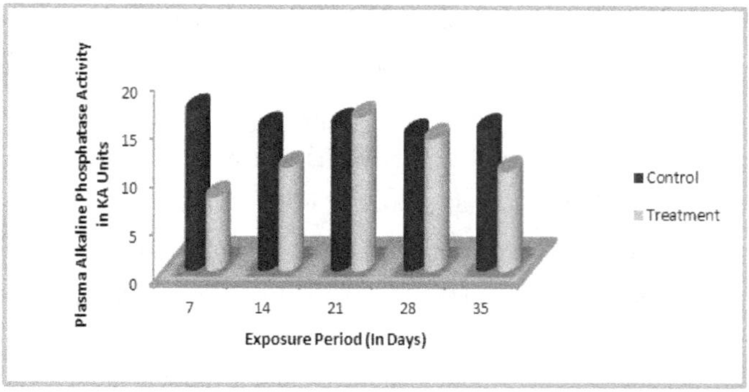

Gill

Observation of gill in control fish

The gill of *Cyprinus carpio var. communis* consists of series of folds, the gill filaments or gill lamellae which are supported by gill rays. The gill lamellae were called primary secondary lamellae. The primary gill lamellae are flat, leaf like, laterally compressed structures and arranged alternatively on either side of the intra bronchial septum. The primary lamellae are lined by inter lamellar epithelial cells. The tips of the primary lamellae are broad containing a dense mass of healthy red blood cells. The gill rays are partly bony and partly cartilagenous and are connected with the gill arch and with one another by fibrous ligaments. The primary lamellae bear on either side, a series of secondary lamellae perpendicular to its long axis. Secondary lamellae of two adjacent gill filaments inter digitized to some extent, so that water passed through small channels. Each secondary lamella consisted of a central vascular layer supported by pillar cells and outer layers of epithelium for respiratory purposes (Plate - II).

Observation of gill in *Cyprinus carpio var. communis* exposure to Aluminum after 28 days

The architecture of gill of *Cyprinus carpio var. communis* shows the marked deterioration. Vacuoles are noticed in the primary lamellae. The cartilaginous rods in the primary lamellae underwent hyperplasia. The shapes of secondary lamellae are not distinct. It revealed disorganized pillar cells and red blood cells. The secondary lamellae are found to be reduced in size, disorganized in structure during the exposure period for 28 days. The proximal region of the gill rays gets damaged and become blunt and swollen. It also possesses disorganized haematomas. The respiratory epithelial cells are found destroyed in most of secondary lamellae. Extensive thinning and destruction of the basement membrane of respiratory epithelial cells is also observed in the secondary lamellae. Shrunken nuclei and vacuoles in the cytoplasm are also noticed in the respiratory cells. Fusions of secondary lamellae are also observed in some regions. Distortion, curling and clubbing in the tips of the secondary lamellae is also observed during toxicity study after 28 days.

Gill Histology

LEGEND FOR FIGURES - PLATE -II

Fig. A. Normal Gill lamella of *Cyprinus carpio var. communis*,
Central axis
Primary gill lamella
Secondary gill lamella
Inter lamellar region
Pillar cell
Erythrocyte

Fig. B. Gill lamella of *Cyprinus carpio var. communis* Aluminum **exposed for 28days**
Epithelial lifting
Bulging of tips of primary gill filaments
Curling of secondary gill filaments
Degenerated secondary lamella

Fig. C. Gill lamella of *Cyprinus carpio var. communis* Aluminum **exposed for 28 days**
Epithelial lifting
Curling of secondary gill filaments
Degenerated secondary lamella
Degeneration of epithelial cells
Fusion of secondary gill filaments

Fig. D. Gill lamella of *Cyprinus carpio var. communis* Aluminum **exposed for 28 days**
Epithelial lifting
Atrophy of secondary lamella
Degenerated secondary lamella
Fusion of secondary gill filaments
Degeneration of epithelial cells

PLATE - II

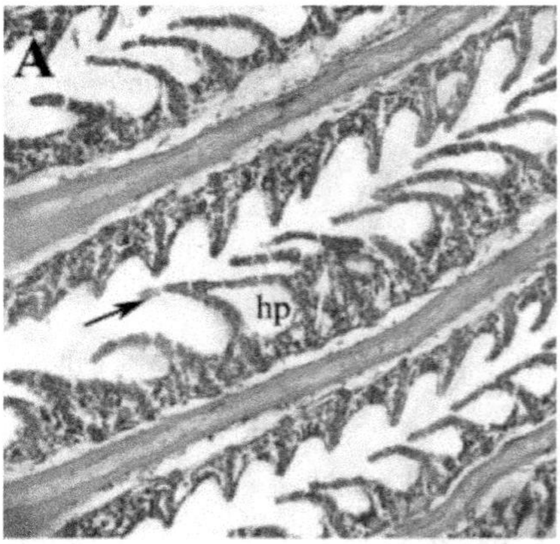

A. Normal Gill lamella of *Cyprinus carpio var. communis*, HE x400

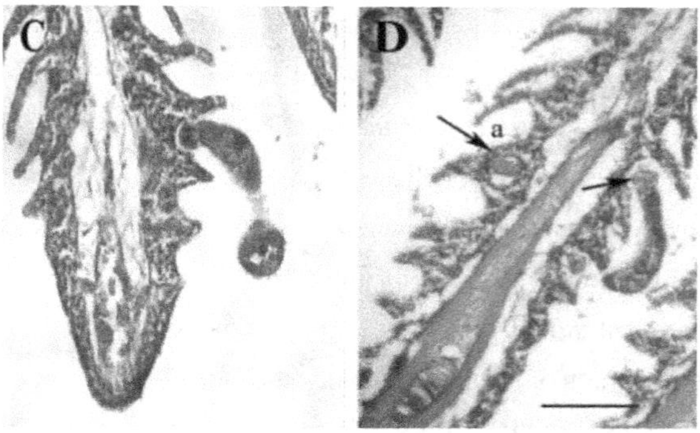

B, C and D- Gill lamella of *Cyprinus carpio var. communis* Aluminum **exposed for 28 days**

DISCUSSION

Bioassay container is distinct as make use of a source of revenue organism to determine the absorption of a material inside water next to formative its strength in produce some definite effects inside the living organism (Alabaster and Lloyd, 1985).

Ferrando also Andreu (1991), report that stationary acute toxicity tests make available rapid in addition to reproducible concentration-response curve used for estimate poisonous belongings of chemical scheduled aquatic organisms which might figure a statistics support for influential relation toxicity in the direction of a mixture of group.

The Median lethal concentration of aluminium sulphate in the direction of *Cyprinus carpio var.* communis used for 24 hours inside the in attendance examination was resolute as 44.90 ppm in addition to used for 96 hours 26.42 ppm. The minority authors have single-minded LC_{50} of aluminium through revealing fishes on neutral pH. Thomsen *et al.,* (1988) calculate the LC_{50} since 3.80 mg Al/L in yielding water in addition to 71.00 mg Al/L into inflexible water into 25 days old larvae of *Salmo gairdneri* by pH 7.0. Also Birge *et al.,* (1980) perform embryo larval bio-assay going on 11 suggestion metals of coal ravage effluents inside embryo of rainbow trout, *Salmo gairdneri* largemouth bass, *Micropterus salmoides* in addition to the streaked salamander, *Ambystoma opacum,* by pH 7.2-7.8. The LC_{50} cost of Aluminium considered intended for trout, bass also salamanders be 0.56, 0.17 plus 2.28 ppm, in that order. Burrows (1977), have in addition particular an explanation of deadly absorption of Aluminium compound in the direction of different species of freshwater fishes. The LC_{50} worth of Aluminium chloride in the direction of rainbow trout, *salmogairdneri* used for 96 hours was 7.4 also 14.6 ppm by pH 6.5 also 7.5, respectively (Call *et al.,* 1984).

Mainly metals in addition to their salt are easy inert compound, the toxicity of which is cause in anions, cations or physicochemical property of the salted; compound of metals contain an unfavorable consequence scheduled the self-purification procedure of a water organization; the dangerous consequence of salt of metals be manifest because form precipitate inexplicable hydroxides of metals deposit on top of the gills in addition to causing humanity of fish, dropping the pH of water on top of hydrolysis also specially impart toxic result base on definite achievement (Metelev *et al.,* 1983).

Aluminium toxicity depends going on the variety of aluminium present, which be mainly independent on top of water pH (Burrows, 1977) in addition to the attendance of complexing ligands such because fluoride (Driscoll *et al.,* 1980), Silicic acid (Birchall *et al.,* 1989) in addition to whole material such because humic acidic (Gundersen *et al.,* 1994).

Two theory contain been projected through the preceding human resources to give details aluminium toxic mode of achievement (Sadler in addition to Lynam, 1987; Handy plus Eddy, 1989). The primary one be material effect in addition to the subsequent one is exterior required. During the original speculation irritation of gill epithelial cells strength comprises be cause by the rainfall of aluminium hydroxide on the gill plane. During the subsequent single it was postulate to facilitate the hydrolysis harvest of Al viz., $Al(OH)_2^+$, $Al(OH)^{2+}$ might attach in the direction of useful group by the gill epithelium important in the direction of casing arrangement and purpose (Sadler also Lynam, 1987).

Some author contain experiential comparable symptom of acute toxicity similar to swelling, edema in addition to swelling also into various cases lamellar combination of the gills (Schofield in addition to Trojnar 1980; Tietge *et al.,* 1988). Into the there study and, behind 24 hours' contact important structural alternations similar to edema also lamellar fusion of the gills be experiential which strength contain lead in the direction of humanity of fish.

Plentiful discharge of mucous in fish on low pH lying on aluminium exposure might block the put in the ground lamellar sitting room into the gill epithelia, most important in the direction of summary water flow in excess of the respiratory surface in addition to greater than before width of the dispersal fence for gas (Westfall, 1945; Ultsch also Gros, (1979). Youson in addition to Neville (1987) experiential aluminium deposit resting on the gills impairing the gas substitute crossways the gills cause cellular anoxia, asphyxiation also acute loss of fish. Conversely, Exley *et al.,* (1996) show to aluminium precipitate on the gill exterior was not constantly connected by an acute toxicity fairly, the might of organization of aluminium by means of the gill epithelium happening during the configuration of hydroxy and oxo-bridges by means of oxygen base group present into the mucous also gill be the influential part.

Al^{3+}, a main aluminium ion by pH < 4.5 be concerned used for acute aluminium toxicity into fish (Erichson in addition to Jones, 1939; Exley *et al.,* (1991) established to the compulsory of Al^{3+} by the gill epithelium be a pre-requisite used for the delicate toxicity.

It is report to $A1^{3+}$ be less poisonous than some additional species of aluminium (Baker in addition to Schofield, 1980; NeVille and Campbell, 1988). But, Playle and Wood (1990) confirmed to Al^{3+} spring to the gill exterior can experience deprotonations in the direction of variety aluminium hydroxide polymers payable to change during the pH of the gill microenvironment also cause toxicity on the way to catch fish.

Some authors contain report to highest toxicity occur by about pH 5.0 to 5.5 everywhere, aluminium hydroxide constitute the main variety (Sadler in addition Lynam, 1987). Then aluminium polymers be recognized because the main reason of acute aluminium toxicity into fish during the pH section 5.0 in the direction of 6.0 (Witters *et al.,* 1996), contradict before observations attribute toxicity in the direction of monomeric variety of aluminium hydroxides.

Aluminium polymerization might start as presently as the $Al(OH)_2^+$ start to yield far above the ground molecular variety of aluminium (Playle also Wood (1990) stated to the gills might act the same as nucleating surface in the lead which aluminium can polymerize. Lydersen *et al.,* (1991) start to the degree of continuing polymerization is a significant part into determining acute toxicity of aluminium. Additional the author's report to fish within call by newly arranged solution of aluminium be showing in the direction of a variety of hydrated aluminium variety, together monomeric in addition to polymeric, solution be greatly poisonous than matured solution.

Newly, Exley *et al.,* (1991) contain explain individuals phospholipids, a membrane transfer protein also polyanionic mucous contain harmfully charge glycoproteins with sialic acid by the gill surface offer possible compulsory site used for aluminium. The constancy of necessary is independent going on the relation concentration of challenging cations also anions in addition to natural binding strong point of aluminium in the direction of gill plus mucous ligands anywhere compulsory can be real weak electrostatic or many management in the direction of a particular aluminium ion (Birchall and Chappell, 1988).

Additional by this pH the main variety of aluminium in attendance into the solution can be present monomeric aluminium hydroxide, which may contain in progress form high molecular polymeric variety. Development of aluminium polymers going on the gill surface furthermore improved mucous discharge strength contain cause severe congestion of the inter-lamellar places primary in the direction of acute hypoxia also humanity of *Cyprinus carpio var. communis,* into the current learning behind the clarification of Zaug (1982), moreover Lydersen *et al.,* (1991) going on the method of acute toxicity into fish.

The ruling of cations similar to Na^{+}, K^{+}, Ca^{+} also Mg^{2+} be disrupt inside water animals below contaminated environment at the same time as report in Bruin (1976). Alteration into the ionic sense of balance of aquatic organisms be able to be real suitable toward stressor effect going on the ion regulatory organs (Gilles also Requeux, 1983) or inside furthermore exterior sensory receptors complex by means of finding of change within osmotic condition (Rosseland plus Staurnes, 1994) or scheduled the endocrine organism (Harman *et al.,* 1980), or scheduled the metabolism (Greenway, 1979), or going on the lively transportation process (Staurne *et al.,* 1984).

McDonald *et al.,* (1989) stated to branchial ionoregulation be the the majority probable physiological aim of surface lively metals suitable in the direction of the attachment range of possible targets also toxic mechanism. Additional the author stated to the possible physiological achieve of metals be scheduled the ionic instruction of fish in addition to it can be suitable in the direction of some of the follow: (i). Disturbance of single or more of the electrolyte transportation mechanism (ii).Widening or blockade of the distribution pathway also iii). Interference by the exterior necessary of calcium by means of an attendant consequence scheduled casing constancy.

Acute contact of freshwater fish in the direction of metals generally results into decrease plasma ions level (Lock *et al.,* 1981). Some authors contain meaningful out to cadmium induce plasma ionic instability be in fact cause in impair uptake as well as diffusional sufferers, of ion by means of the gills also significant toward hyponatremia, hypochlorenemia also hyperkalemia (Pratap *et al.,* 1989).

According in the direction of Poleo plus Muniz (1993), an improved width of the distribution difficulty among water as well as plasma suitable toward aluminium polymerisation also mucus clearance can decrease the accessibility of electrolytes used for the transfer molecules integrated during the gill epithelial covering. This find maintain to ion arrival might be present artificial in an swelling plus edema of the transport gill epithelia into response toward surfacial aluminium make up at home brown trout, *Salmo trutta,* because report in Karlsson-Norrgren *et al.,* (1986a,).

Freda *et al.,* (1991) contain suggested to equally the dislocation of casing spring Ca^{2+} by sub-sequent disadvantage of the tense junction through deformations along with stimulating of the gill epithelium can contain accelerate ion loss less than Al stress. The higher than method of aluminium into depress plasma Na^{+} also Cl^{-} concentration be support in Vuorinen *et al.,* (1992).

Into the current study the defeat of Na^{+} as well as Cl^{-} ions into acute along with sub-lethal study might be suitable toward the impair uptake also diffusional sufferers *via,* gills into *Cyprinus carpio var. communis* by means of aluminium stress because recommended by the higher than authors. Larsson *et al.,* (1976) opined to osmoregulatory disappointment capacity be a explanation for decrease level of main plasma electrolytes. McCarty also Houstan (1976) stated to the reduce into plasma electrolytes levels tend near join with raise within tissue concentration.

Caldwell (1977) celebrated to the reserve of ATPase activity during gills be a demonstration of the mutilation of the ionic balance along with rule. Staurnes *et al.,* (1984) confirmed to ion entry within rainbow trout, *Salmo gairdneri,* might result in a straight line starting reserve of transfer enzymes like Na^{+}, K^{+}, and ATPase by means of aluminium within the gills. The reduce into chloride uptake in aluminium showing salmonids might occupy embarrassment of HCO_3^{-} ATPase the same as opined by means of Battaram (1988). A like device *i.e.,* reserve of ATPase action in aluminium toxicant might be working in the current study.

NeVille in addition to Campbell (1988) experimental rainbow trout, *Oncorhynchus mykiss,* treat by means of aluminium by pH 5.5 to 5.0 show reasonable ion flux response furthermore a alteration among the two main retort mechanism inside fishes (Respiratory in addition to ionoregulatory disturbances).This bimodal physiological reaction be and experimental by means of Poleo *et al.,* (1995) inside fish. A like, bimodal device might be in service into the current study all through acute aluminium exposure because indicate by reasonable reduce into Na^+ plus Cl^- level during the plasma of *Cyprinus carpio var. communis,* everywhere the pH of the investigational water was concentrated to 5.0 + 1 behind accumulation of LC_{50} 24 hours' attention of aluminium sulphate.

Bigger level of ions within the plasma of fish into sub-lethal management might be recognized in the direction of an improved mixture of move proteins in charge used for sodium uptake (McDonald and Milligan, 1988) or in rising gill compulsory affinity in support of Ca^+ so offset the ready for action necessary of Al^{3+} (Reid *et al.,* 1991) or in the propagation of "Chloride cells" going on the branchial epithelium of fully developed salmonids (Laurent and Hebibi, 1988).

Booth *et al.,* (1988) include report used for Na^+ raise, to the response of native fish in the direction of acidic aluminium contact might be characterised since happening inside two phases (i). The 'shock phase', (0-48 hours) recognized by substantial efflux of sodium plus chloride in addition to reserve of sodium uptake on the gill. (ii).The upturn phase' anywhere efflux be concentrated regardless of continuous reserve of uptake.

During agreement by means of the study, into the current study all through sub-lethal contact, originally Na^+ level decrease behind 1st week signifying 'Shock phase' furthermore subsequently improved into the succeeding weeks exhibit 'Recovery phase'.

According in the direction of Heming and Blumhagen (1988), the experimental K^+ ion increase during the plasma of rainbow trout, *Oncorhynchus mykiss,* all through Acid moreover aluminium contact, suggest the descent of cell membranes by means of resulting leakage of cellular potassium interested within the blood plasma. Increase into the plasma K^+ might also be alive due in the direction of decrease during the extra-cellular space (Witters, 1986) or decrease plasma capacity (Playle *et al.,* 1989). A like circumstance into the current study might account used for the raise into plasma K^+ level recall the more than clarification.

Within the current study, blood pH plus bi-carbonate levels decrease within acute management. But, no important alteration was experimental into the sub-lethal study. Single of the the majority essential tasks for homeostatic ruling is the preservation of a set position pH into the body fluid of an animal (Cameron, 1976). Wood *et al.,* (1984) include indicate with the intention of the exchange charge of Na^+, H^+, (NH4) and Cl^-, HCO_3^-, $(OH)^-$ are vigorously manipulate inside an correct comportment for the duration of acid-base instability to complete inside pH reimbursement.

Bass plus Heath (1977) recommended to the reserve of oxygen uptake (Hypoxia) during the gills bring a transfer toward anaerobic metabolism cause decline into blood pH. According toward Sellers et *al.,* (1975), zinc publicity inhibits gaseous exchange inside the gills of fish, in that case along with major blood pH decrease, *i.*e., lacto acidosis occurs probably due to the accumulation of lactic acid in addition to CO_2 during the blood. Aluminium cause decreases into blood pH, which was credited in the direction of respiratory acidosis, appropriate to CO_2 addition or to metabolic acidosis appropriate toward lactic acid gathering at the same time as a result of anaerobic respiration into rainbow trout, *Oncorhynchus mykiss,* all through acid also aluminium exposure (Playle *et al.,* 1989).

Low pH environment be possible in the direction of get in the way with acid-base rule during gills within two significant habits. Initially such atmosphere inhibit lively ion transfer troublesome Na^+/ H^+ in addition to Cl^-/ HCO_3^- replace plus thus disable rapid response of fish on the way to blood acid-base turbulence (Heisler, 1982). Although kidney remainder reasonably competent into acid-base guideline by low pH (McDonald plus Wood, 1981), it cannot completely return the meaning of gills because it respond extremely gradually to acid confront (Wood as well as Caldwell, 1978). Secondly, the gills container act the same as a main route used for H^+ ion access of body fluids as soon as peripheral H^+ be important (McDonald, 1983b).

Additional H^+ ion shaped all through aluminium hydrolysis or aluminium induce hypoxia consequential into anaerobic metabolism or impair gaseous exchange appropriate toward mucus difficulty by the gill surface irritating lactic acid in addition PCO_2 accretion also reserve of acid-base related ion exchanges strength be the possible cause of blood pH refuse experimental in the current study.

According in the direction of Staurnes *et al.,* (1993) the refuse of HCO_3^- stage into the plasma of *Oncorhynchus mykiss* be suitable to reserve of carbonic anhydrase enzyme by means of aluminium. A comparable motive was qualified toward blood HCO_3^- reduction inside trout, appropriate to nitrite publicity in Williams also Eddy (1988). A different clarification presented by means of Houston plus Mearow (1982) used for the refuse of bicarbonate be as maintain plasma electro objectivity single or more anions with bicarbonate might fall.

Some metals impose anaemia into fishes (Gill *et al.,* (1991) report to anaemia accompany by decrease into haematocrit, haemoglobin also red blood cells be the major haematological consequence of metal intoxication within fish.

Anaemia can be suitable in the direction of a raise of plasma capacity cause with a disrupt water balance, decrease velocity of RBC manufacture, defeat or obliteration of red blood cells, because recommended by means of Larsson *et al.,* (1976). Further mechanism disturbed by anaemia might contain power of metal scheduled the life span of erythrocytes (Berlin in addition to Friberg, 1960)

Narain plus Srivastava (1989), optional to the low haemoglobin level into strained *Oncorhynchus kisutch* also *Heteropneustes fossilis* would be present an estimated penalty of red blood cell failure. The author's additional supplementary to insufficient haemoglobinisation or reduction of red blood cells is able to and be present a source used for decrease haemoglobin in addition to haematocrit value.

Pamila *et al.,* (1990) into *Sarotherodon mossambicus* in addition to Ghazaly (1992) into *Tilapia zilli* report the important lesser comfortable of haemoglobin in cadmium treat fish than organize values signifying the contaminated achievement of cadmium scheduled haemometabolism. Dalwani *et al*., (1985) contain describe the communication involving metals plus enzymes concerned within haemometabolism.

Erythropenia as well as an associated decrease into haemoglobin also haematocrit levels are the demonstration into fish, *Barbus conchonius*, experimental in Gill plus Pant (1987) below chromium toxicities. The author's declared to morphological aberration all through erythrocytes cause in determined publicity in the direction of chromium resulted into cheap oxyphoretic ability of RBC plus consequently leads into the way of tissue hypoxia; while erythropoiesis is inspired at some stage in hypoxia (Grant with Root, 1952), appropriate to expanded contact of chromium, erythropoiesis fail near maintain rapidity by the physiological demand. Additional the author explain to organal damage cause in chromium in this fish, counting passageway obstruction of the secondary lamellae plus telangiectasis and renal lesions might give considerably on the way to secondary anaemia.

During conformity toward this view Gill *et al.,* (1991), also opined to erythroclasia (RBC destruction) surpass the red blood cells manufacture plus release as well as the determination of poisonous cations dominate any compensatory consequence of the showing fish. Metal cause haemolysis plus reserve of erythropoiesis be report in some author (Sakthivel, 1988).

Here the current study and, aluminium showing fishes exhibit anaemic provision in recording minor substance of haemoglobin, haematocrit also red blood cells than the manage fish. This might be outstanding to construction of RBC or reserve of erythropoiesis or interference of the hemometabolism or trouble in fluid capacity sense of balance formed by means of aluminium ions which might find maintain from the higher than employees.

The reduce of haemoglobin lead in the direction of hypochromic microcytic anaemia, which be recognized toward insufficiency of iron also its decrease consumption used for haemoglobin mixture; It be healthy recognized to glycolysis be worried by the decrease of met-haemoglobin since almost immediately because it is produced, thus maintain the iron of the haemoglobin into the ferrous form, inside which position only, it act the same as an professional O_2 transporter (Bhai *et al.,* 1977). Koundinya with Ramamoorthy (1977) be of the estimation to the disturbance of iron synthesize technology suitable to reserve of aerobic glycolysis might be the motivation used for the reduce of blood parameters into the strained a fish. During the current study, a comparable type of method might be in service resulting into the decrease of haemoglobin count within *Cyprinus carpio var. communis.*

Gill with Pant (1987), recorded into *Puntius conchonius* behind mercury publicity and preliminary fall inside RBC in addition to haemoglobin count, although behind 2 or 3 weeks, increase occur in together these index. The author describe to incipient reduce could be present due in the direction of haemolysis cause in mercury, although sub-sequent revival can be recognized to improved erythropoiesis which be trigger because characteristic pressure reaction. Inside contract in the direction of this impression, Witters et *al.,* (1990) report to increasing catecholamine concentration inspired spleenic reduction to raise the quantity of circulate erythrocytes.

Saad *et al.,* (1973) report to into fishes, blood constituent such since haemoglobin, RBC plus others contain rapidly improved in the strain of hypoxia. Within our current sub-lethal study, the occurrence of steady upturn in RBC in addition to haemoglobin might be connected to improved erythropoiesis because a stress reply or hypoxia induce improvement as report by means of the more than authors.

Folmar (1993), declared that increase into haematocrit value be linked by means of osmotic shifts; the same as the pH of the blood decrease, erythrocytes swell in addition to plasma amount decrease, the conclusion of Hilmy *et al.,* (1987) maintain the higher than analysis, to RBC inflammation be the motive used for haematocrit raise in fishes behind contact in the direction of chromium, copper, mercury also zinc correspondingly.

The more than conception is additional strengthen in the study of high value of haematocrit into fishes in aluminium exposure into acidified waters (Reid *et al.,* 1991). Upper values of haematocrit experimental into the current study behind 7th plus 35th week might discover maintain of the on top of author's clarification to the RBC inflammation consequential to liquid volume trouble strength contain improved the haematocrit value.

Leucocytosis experimental in the toxic produce of manganese into *Heteropneustes fossilis* be an alteration to get collectively the stress situation in fish since reported in Garg *et al.,* (1989), declared to oxygen insufficiency is answerable used for profusion of leucocytes. Singh (1995) explain to leucocytosis be almost certainly due in the direction of the tissue injure in addition to sub-sequent elimination of cell fragments in Cu also Cr contact in *Channa punctatus.* Nelson in addition to Rani (1992) contains and supports this analysis by means of the actuality to toxicant damage gill tissues as well as purpose, leading in the direction of leucocytosis. Lymphocytosis be experimental in Chitra plus Ramana Rao (1986) into *Channa punctatus* by means of mercury intoxication, which stimulate lymphopoiesis because a straight stimulus of the immunological method in opposition to stressor current in the medium.

Within line among more than author's results, into the current study in addition, leucocytosis be observed into both acute plus sub-lethal exposure of aluminium in the direction of meet stress situation which strength contain formed hypoxia also gill injure within fish.

Ruparelia *et al.,* (1991) recommended to red blood cell inflammation is the origin used for the reduce within MCHC value into *Oreochromis mossambicus* behind cadmium experience. Nikinmaa (1982) in addition to Vuorinen *et al.,* (1992) report to lower MCHC, important haematocrit denote the enlargement of erythrocytes, a method in which blood O_2 transportation ability is improved while fishes be subjected in the direction of a smaller amount effectual gas exchange (Vuorinen *et al.,* 1992). In addition lowered MCHC strength result starting the discharge of youthful erythrocytes contains less haemoglobin than the big ones (Lehmann and Sturenberg, 1975) interested in the transmission (Vuorinen *et al.,* 1990).

Here support of the higher than concept various investigators record refuse into MCHC worth *viz.,* by means of sensational aluminium in the direction of fish into acidified water (Vuorinen *et al.,* 1992). During the current study the reduce into MCHC significance might be suitable to red blood cell enlargement or let go of youthful erythrocytes contain a lesser amount of haemoglobin keen on the transmission.

Ghazaly (1992), report decrease within MCH value inside *Tilapia zillii* behind cadmium poisoning. Additional the author confirmed to these haematological change into fish reimburse for impair oxygen uptake suitable in the direction of gill injure cause in cadmium. At home the current study, comparable haematological change might contain occur due in the direction of aluminium publicity.

The bigger Mean Cell Volume (MCV), all through strain (Capture) is due in the direction of the inflammation of red blood cells or toward the discharge of big red blood cells interested during the common movement as report by means of Fletcher (1975). The anaerobic situation (Hypoxia) almost certainly residential all through sub-lethal aluminium strain (Poleo, 1995) might be present the motivation for improved Mean cell volume because indicates in Soivio plus Nyholm (1973) to anaerobic storage of trout red blood cells cause them in the direction of swell up.

Vallee with Ulmer (1972), report to metals explain a physically powerful similarity used for ligands such because phosphates, cystenyl in addition to histidyl surface sequence of proteins, purines in addition to porphyrins be capable of form complex by ligands contain sulphur, nitrogen also oxygen because electron donor.

Goodman in addition to Gillman (1970), experimental to the majority toxicants be transported reversibly jump in the direction of blood protein by simply little section existing into the free of charge or liberated situation. Toxicants should regularly be real into the free of charge state toward respond by natural molecules plus get in the way by means of biochemical mechanism. The writer in addition emphasize the require of the plasma protein accessibility in favor of the compulsory plus transport of deadly substance which might play a significant role during the toxicological penalty of these compound.

Total body protein concentration be partial in a mixture of ecological factor (Claybrook, 1983). Below situation of strain various organisms determination organize proteins since an power starting place (Bayne, 1973a,) fast consumption of protein in the fishes below contaminant strain have be report in Palanichamy also Baskaran (1995).

Srivastava and Srivastava (1984) be of the estimation to in stress, fish might option to substitute metabolic pathway mediate by means of a decrease in the movement of Nissl also Golgi granules. Neff (1985) declared to decline within protein contented might be connected to impair provisions intake, improved energy cost used for homeostatis, tissue restore plus detoxification mechanism for the period of strain. Decrease into protein synthesis have and be reported in the direction of result starting the reserve of development of latest peptide chains in addition to the dissociation of 80S ribosome since propound by means of Rao *et al.,* (1987).

Chandravathy with Reddy (1994), declared to reduce into protein strength involve poverty of protein interested in amino acids cause an increase of amino acid collection. The amino acids additional can be use for gluconeogenetic purpose. Such reports contain be recognized by some author into fishes below toxicant strain (Chandravathy also Reddy, 1994). Diminish incorporation of amino acids interested in the protein in addition capacity lead in the direction of reduce into protein comfortable the same as opined in Dhar also Banerjee (1983).

The reduction of protein satisfied can propose improved proteolysis in addition to potential use of the harvest of their degradation used for metabolic purpose. The harvest might be fed interested in TCA cycle all the way through amino transferase method to satisfy the overload require of energy for the duration of toxic strain circumstances as experimental into freshwater fish *Anabas scandens* behind contact in the direction of lead nitrate (Chandravathy plus Reddy, 1994).

Improved adrenocorticosteroid yield by means of nickel contact within *Cyprinus carpio* (Thatheyus *et al.,* 1992) be connected by protein reduction. Ghosh in addition to Chatterjee (1989) optional to protein reduction in toxicant exposure be possibly suitable in the direction of (1). Straight use by cells in support of power supplies, (2). Development of lipoproteins which be utilize for repair of injured cells as well as tissue organelles also (3). Improved lipolysis. The reduce in protein content in the current study for the period of acute in addition to sub-lethal treatment by aluminium might be suitable to rapid consumption of protein in fishes, bigger proteolysis or cheap protein synthesis below aluminium strain recall the clarification of the higher than author's.

However, into the current study behind 2nd and 3rd weeks of sub-lethal exposure in the direction of aluminium, plasma protein satisfied of *Cyprinus carpio var. communis,* show improvement. This might be outstanding in the direction of an raise into the charge of protein synthesis because a 'Common Adaptation Syndrome' toward join the excess aluminium ion into the plasma, since aluminium have a very high similarity in support of a big quantity of proteins, poly-nucleotides in addition to glycosamine glycans plus probably binds by phosphorylated proteins also phosphorylated histones (Martin, 1994). This analysis find maintain starting the statement of Ramalingam with Ramalingam (1982).

Bouck as well as Ball (1965) in addition established to small O_2 concentration changed the protein work of serum into blue gill, *Lepomis macrochirus* plus largemouth bass, *Micropterus salmoides.* Bouck (1972), establish into rockbass, *Ambloplites repestris* undergo sub-lethal hypoxia, an important raise inside plasma protein, which be recognized in the direction of cellular enzymes entering the blood. Hypoxia, generally experimental behind aluminium exposure into fishes (Poleo, 1995), strength contain in addition contribute in support of the raise into plasma protein content within the current study.

Rosseland plus Staurnes (1994) describe to major consequences causal toward the casualty of fish for the duration of acute aluminium exposure absorption be respiratory, ionoregulatory in addition to circulatory instability. Into accumulation, in the direction of the more than characteristic symptom, improved blood protein absorption too leads toward improved blood thickness. Subsequent to ion disturbance, the protein plus amino acids contribute actively as they are osmotic effectors because fine because mechanism into branchial buffer system (Jurss, 1980). During the current study for the duration of sub-lethal management the short-term elevation into protein content going on days 14th plus 21st might contain act because an osmoeffector element consequent toward incidence of ion conflict created by means of aluminium stress within *Cyprinus carpio var. communis.*

An important blood glucose level be a common reaction below most stressors (Waring also Brown, 1997). Xenobiotics of varied origin formed hyperglycaemia with hepatic glycogen depletion into different teleost species (Sivakumari, 1998).

Pickering (1993), report to the hypothalamo-pituitary interrenal axis, activate in numerous stressors, important blood levels of cortisol. This endocrine response be precede in a surge of catecholamine's which stimulate lipolysis, glycogenolysis in addition to gluconeogenesis which be significant power mobilize process (Vijayan and Moon, 1992).

Starting the review of Folmar (1993), it be evident to a close connection exists among the raise into serum cortisol plus glucose level. The author contain additional report to the rapid rise into glucose results starting glycogenolysis (Release of glycogen reserves into muscle also liver) initiate in catecholamines as continued elevation of serum glucose be maintain by means of cortisol inspired gluconeogenesis (Protein metabolism). A related rise into glucose level simultaneous by means of increase of plasma cortisol is investigational behind aluminium exposure in the direction of a mixture of variety of fishes (Waring also Brown, 1997).

Goss with Wood (1988) record an apparent raise into blood glucose level into rainbow trout, *Oncorhynchus mykiss,* below acute aluminium into acidic waters. The authors contain record a simultaneous raise into plasma cortisol along with better ionoregulatory trouble. Results maintain the phenomena to plasma cortisol mediate glycogenolysis in addition to the resulting hyperglycaemia aid into instruction of plasma osmolarity. Under circumstances of strain, hyperglycaemia might present extra energy for the duration times of metabolic wants such because fight or flight reaction (Goss and Wood, 1988).

Metals might adjust enzyme activity interested in animals within some ways (Kench, 1972; Verma *et al.,* 1983). They might bind happening the technique in the direction of macromolecules during the cell frequently by means of enzymes (Britten plus Blank, 1973) or shift the absorption of co-factors or reactants through varying film permeability in addition indirectly disturbing enzyme actions (Tucker also Matte, 1980).

Na^+, K^+, ATPase have be reviewed as well as assessed since a potentially helpful pointer of pollution pressure into aquatic animals (Torreblanca *et al.,* 1989). Reserve or stimulus of Na^+, K^+, also ATPase activity might be real expected near contain metabolic or ionic achieve within fishes into relation in the direction of osmoregulation because stated by means of Verma *et al.,* (1983).

During agreement by means of this, copper decrease the Na^+, K^+, ATPase action into the gills of *Coho salmon, Oncorhynchus kisutch,* (Lorz plus McPherson, 1976). Na^+, K^+, ATPase action was establish in the direction of be reserved inside Chinook salmon, *Oncorhynchus ashawytscha,* in copper exposure since reported in Beckman in addition to Zaugg (1988). The author's recommended to copper damages 'chloride cell' somewhere Na^+, K^+, ATPase is placed also impaired its function plus the enzyme be inactivated by means of the straight reserve of metals.

According on the way to Panchalingam *et al.,* (1991) aluminium toxicity is articulated in its facility on the way to join phosphate, ATP, or an additional functionally connected organic compound contain phosphorus. Therefor ATP forms stronger complex by AL^{3+} than among Mg^{2+} also *the* Al^{3+} ATP complex act because a strong inhibitor on the way to Mg^{2+} ATP like indicate in Ganrot (1986).

The opportunity of straight reserve of enzyme in aluminium in addition exists since Al have been shown on the way to accumulate into the chloride cells of together brown trout, *Salmo trutta* with rainbow trout, *Oncorhynchus mykiss,* while kept into Al wealthy acid water (Youson also Neville, 1987). During brook trout, *Salvelinus fontinalis,* aluminium have been established toward increase speed the proceeds plus regeneration of the chloride cells (Chevalier *et al.,* 1985). As the high gill action of Na^+, K^+, ATPase into smolts of Atlantic salmon, *Salmo salar,* is localised during the chloride cells, such degeneration of these cells might have contribute towards the small gill Na^+, K^+, ATPase action of the aluminium showing fish (Staurnes *et al.,* 1993).

Into the current study, the reserve of Na^+, K^+, ATPase activity into the gills of fish *Cyprinus carpio var. communis* might be present suitable to straight reserve of enzyme or replacement of common co-factor Mg^{2+} ions or 'Chloride cell' injure or in binding of the metal ion among the phosphate groups of enzyme recall the clarification of the higher than author's.

The raise into Na^+, K^+, plus ATPase activity might be suitable toward 'hormesis' within fish since pointed out in Yang also Randall (1996). The author's define the time 'hormesis' since an over reimbursement in the direction of several inhibitory challenge. Thurberg *et al.,* (1977) noted a statistically important raise within ATPase activity in addition to gill tissue respiration into lobsters, *Homarus americanus,* showing on the way to cadmium which might be appropriate to improved metabolic activity in gill tissues, rather than organism a straight effect of cadmium scheduled the enzyme.

The xenobiotic preserve interact straight by the enzyme or change Na^+, K^+, in addition to ATPase activity appropriate on the way to disturbance of power produce metabolic pathways (Yadwad *et al.,* 1990). ATP provisions are cheap by improved energy request into response on the way to a physiologically stressful situation. While Na^+, K^+, ATPase turnover rate be really rapid, Na^+, K^+, ATPase would be present responsive on the way to factor disturbing protein biosynthetic pathway (Haya also Waiwood, 1983). A tall proceeds of ATPase within plasma into the current study seeing that evident starting the peak action occurred into the middle of the publicity period, was related by decrease plasma protein.

Chemical in addition in the direction of non-chemical stressors preserve lead on the way to release of cellular enzymes probably appropriate to localised ischemia, reduction of nucleotides in addition to tall energy phosphates or change within membrane possible (Rattner *et al.,* 1983). Metals contain and are shown on the way to decrease membrane solidity probably most important on the way to enzyme leakage (Versteeg plus Giesy 1986). improved serum lysosomal enzymes is an suggestion of toxicity which might be suitable to improved enzyme manufacture, decrease lysosomal strength or tissues injure because opined by means of Mayer *et al.,* (1992).

According on the way to Ganrot, (1986) inside cells, Al^{3+} accumulate primary into the lysosomes in addition to subsequently be gradually transfer on the way to the nucleus of the cell. The writer promote report to the toxicity effect might be level greater by a additional neutral pH but Al^{3+} have satisfactory time toward bind during sufficient quantity in the direction of the membranes.

Lieberherr *et al.,* (1982), show to into human being skeletal tissue acid in addition to alkaline phosphatases be activate by Al^{3+} by lower absorption as well as reserved at advanced concentration. Siegel *et al.,* (1983), report to Al^{3+} bind stoichiometrically with kindly toward calmodulin, the usual activator of the enzyme along with thereby induce a main structural alteration. Ganrot (1986) declared so as to reserve of phosphodiesterase (ALP) be possibly the answer of a far above the ground similarity relations among Al^{3+} furthermore calmodulin. Reserve of alkaline phosphatase action within *Cyprinus carpio var. communis* in sub-lethal publicity by aluminium might enclose be suitable to straight reserve of enzymes which might find maintains starting the clarification of more than employees.

Toxicity impact of aluminium on histological in addition to histopathological study going on the chosen organ of *Cyprinus carpio var. communis*

Gill:

Histopathological changes notice during the gill of *Cyprinus carpio var. communis* showing on the way to Aluminium is suitable to the compassion of the gill tissue plus its respiratory exterior which have straight write to during aquatic environment by means of metal pollutant. Pathological lesion observed during the respiratory epithelial cells of *Cyprinus carpio var. communis* showing to sublethal attention of aluminium. On the publicity of 28 days, the of *Cyprinus carpio var. communis* show like variations into the architecture of the gill, attendance of mucus substance, major gill lamellae demonstrate partial loss of respiratory epithelium plus in addition, the cartilaginous region into the major lamellae show hyperplasia, gill energy become thicken by vacuoles. The minor lamellae are slightly injured in addition to show necrosis furthermore pycnosis occasionally. Also the sub lethal absorption of chromium going on 28th days changed the architecture of gill of *Cyprinus carpio var. communis* by the obvious descent. Vacuoles are notice into the primary also secondary lamellae was established on the way to be decrease in size, disordered within structure. The cartilaginous rod into the major lamellae undergoes hyperplasia by means of the indistinct profile of secondary lamellae. It exposed confused pillar cells plus red blood cells, in the finish of the similar exposure stage. The proximal section of the gill energy was injured in addition to become blunt as well as swollen. It in addition possesses disordered haematomas. The respiratory epithelial cells were established damaged in the majority of minor lamellae. General thinning plus damage of cellar membrane of respiratory epithelial cells were in addition observed within the secondary lamellae. Shrunken nuclei also vacuoles into the cytoplasm were and noticed into the respiratory cell. Fusions of secondary lamellae are and experimental into several region. Alteration curling also clubbing into the tips of the secondary lamellae is and experimental for the duration of toxicity study behind 28 days also it was apparent by means of scanning electron microphotographs.

The clarification made into the current experiment is comparable on the way to the findings of some authors. Histopathological property of sublethal concentration of monocrotophos going on the gills of *Anabas testudineus* is report in Santhakumar *et al.* (2001). Histopathological changes experimental be hemorrhage inside the primary also secondary gill lamellae, disintegration with necrosis of epithelial cells, alteration of the secondary lamellae, disturbance of epithelial cells starting support cells, shorter gill lamellae, synthesis, total destruction of lamella, improved vacuolation, irregular manifestation of gill lamellae be experimental within *Poecilia reticulata* showing in the direction of chlorpyrifos (De Silva plus Samayawardhena, 2002). Degenerative changes into gill, such the same as intraepithelial edema during the secondary lamellae, thick coating of mucus layer resting on the whole gill filament also lamellae, attrition of secondary lamellae, thicken of lamebllae, swelling of epithelial cells, breakages into primary lamellae, disintegration of secondary lamellae, necrosis, break of epithelium were notice for the duration of exposure of sublethal concentration of monocrotophos within *G. affinis*. Experimental uncontrolled restoration of the primary lamellae in addition to secondary lamellae, hypertrophy, hyperplasia, necrosis of the epithelial cell, epithelial stimulating, also dilation of the blood sinuses of the secondary lamellae, lamellar aneurism, moreover hemorrhages during the gill of fingerlings showing on the way to profenofos. Ashok Kumar Gupta also Ashwani Kumar, (2006) contain experimental the mercury ions induce severe histopathological changes inside gill such because degeneration into the middle also distal region into the primary lamellae. Here the distal region globular structure is produced due toward degeneration also fusion into pillar, epithelial plus mucus manufacture gland cells the length of by large number of blood cells of the minor gill lamllae. Muthukumaravel also Rajaraman, (2013) establish so as to the poisonous consequence of chromium scheduled the histology of gill also liver of *Labeo rohita* fingerlings going on different publicity.

Yilmaz *et al.* (2011) experimental the histophathological changes into the *Leuciscus cephalus.* They reported to fingerlings showing in the direction of $CdSO_4$ be exposed hydropic deterioration plus necrosis of the minor lamellar epithelia also inflammation of chloride cells. While related results within Cd publicity be seen into additional studies scheduled dissimilar species of fingerlings, various other histopathological changes be in addition report. Publicity of mercuric chloride preserves reason strict pathological alteration into gills also liver of *Cyprinus mrigala.* The main changes into the liver consist of defeat of cellular structural design, necrosis during hepatocytes also addition of fat inside parenchyma cell. The gills show lamellar collapse, epithelial stimulating also necrotic changes into intercellular epithelial cells (Chavan also Muley, 2014).

SUMMARY

- The 24 hours Median lethal concentration of aluminium sulphate on the way to fish is 44.90 ppm in addition to the equal for 96 hours was 26.42 ppm; representative to the metal is poisonous in the direction of fish.

- Experiments are conducted used for a stage of 35 days among weekly sample regularity. No humanity was experimental during the investigational period. Winning completion of the predetermined revelation periods of 7, 14, 21, 28 as well as 35 days, 20 fishes were taken

- Totaling of Median lethal concentration of aluminium sulphate cheap the pH of the analysis water from 7.5 ± 1 to 4.9 ± 1. During this pH variety polymerisation of monomeric aluminium hydroxide variety to upper molecular variety might happen which be very toxic on the way to fish.

- Statement of toxic aluminium polymers going on the gill exterior strength include cause structural alteration such as edema, lamellar synthesis of gills also plentiful discharge of mucus cause severe congestion of the interlamellar seats. The more than alter strength include resulted into impair gaseous exchange within gills most important on the way to mortality of fish for the duration of acute management.

- Na^+ in addition to Cl^- ions be condensed within the plasma of fish showing to acute also sub-lethal concentration of aluminium. This might include be due on the way to the statement of aluminium polymers going on the gills top in the direction of decrease into ionic replace in favor of transfer, payable on the way to the reserve of active ion arrival, stimulus of inactive efflux during paracellular channel or payable to the reserve of transport enzyme inside gills.

- Elevation into plasma potassium within the current study rnay have been due on the way to split of cell membranes cause discharge of cellular potassium within to the plasma decrease in the extra-cellular gap or reduce inside the plasma capacity.

- Blood pH also HCO_3^- level of fish treat by acute aluminium toxicity decrease considerably. Though into sub-lethal management the higher than change be not important. Surplus H^+ ion formed for the duration of aluminium hydrolysis, aluminium induce hypoxia payable in the direction of impair gaseous exchange irritating lactic acid, PCO_2 addition, inhibition of acid-base significant ion exchange strength be the possible causes of blood pH refuse experimental into the current study. Decline of HCO_3^- level into the plasma inside the current study might include be due in the direction of reserve of carbonic anhydrous enzyme.

- Fish showing on the way to together acute also sub-lethal study exhibit cheap levels of haemoglobin, haematocrit also RBC count than so as to the control. This might have been due on the way to trouble inside the fluid volume balance, reserve of erythropoiesis, erythroclasia or interfering of hemometabolism by means of aluminium ions. Leucocytosis experimental into the acute also sub-lethal treatment is almost certainly due toward stimulated immunological method aligned with aluminium strain. Enlargement of red blood cells might have been the reason in support of the major changes in MCHC also MCV values within sub-lethal study.

- The reduce in the plasma protein content during the present study acute in addition on the way to sub-lethal treatment might have been in the direction of rapid utilisation of protein improved proteolysis, decrease in the protein synthesis in the fish below aluminium strain. All through sub-lethal management hypoxia might have been responsible used for the increase of plasma protein satisfied which might act because an osmoeffector subdivision subsequent in the direction of ion trouble caused in aluminium toxicity.

- Plasma glucose be elevated into both acute also sub-lethal treatment. This might include been due in the direction of improved plasma cortisol level into the investigational fish resultant within glycogenolysis or gluconeogenesis on the way to present extra power for the duration of the times of metabolic strain.

- Gill in addition to plasma Na^+, K^+, ATPase action of fish was reserved in aluminium toxicity both into acute also sub-lethal study. This might include been due in the direction of straight reserve of enzyme in aluminium ions substitute of common co-factor Mg^{2+} ions of the enzyme, compulsory of the metal ion by means of the phosphate grouping of the enzyme or 'Chloride cell' injure. The raise in the plasma Na^+, K^+, ATPase into the sub-lethal action behind 21st days might have been payable to 'hormesis' (Yang plus Randall, 1996) which might be a reimbursement development toward inhibitory challenge created in aluminium.

- Together Acid in addition to alkaline phosphatases actions bigger in the investigational fish which might contain is due in the direction of lysosomal membrane injure cause leakage of cellular enzyme into plasma. But a reduce inside the plasma ALP activity be record into the sub-lethal action. This might be recognized to straight enzyme reserve in aluminium metal which be reported on the way to include a high similarity communication by means of calmodulin, the nonnal activators in the enzyme (Ganrot, 1986).

- The architecture of gill of *Cyprinus carpio var. communis* shows the noticeable descent. Fusions of minor lamellae be and experimental into several region. Alteration curling also clubbing into the tips of the minor lamellae is in addition experimental for the duration of toxicity study behind 28 days.

BIBLIOGRAPHY

Athikesavan, S., Vincent, S., Ambrose, A., & Velmurugan, B. (2006). *Journal of Environmental Biology.* **27: pp** 391-395.

Alabaster, J. S. and Lloyd, R. (1985). *Butterworth Scientific Publications,* London.

Alexander, J. B. and Ingram, G. A. (1980). *J. Fish Biol.* **16:** 115-122.

Apha. (1971). *American Public Health Association,* New York, USA, p. 874.

Apha, Awwa also Wpcf. (1985). *American Public Health Association,* Washington, USA.

Atland, A. (1998). *Environ. Biol. Fish* **53:** 331-345.

Ashok Kumar Gupta also Ashwani Kumar, (2006). *Articles. Journal of Environmental Biology.*

Baker, J. P. and Schofield. C. L (1980). *A.Sandefjord, Norway, SNSF,* pp. 292-293.

Bass, M. L. and Heath, A. G. (1977). *Water Res.* **11(6):** 497-502.

Bayne, B. L. (1973a). *J. Mar. Bioi. Assoc. U. K.* **53:** 39-58.

Battaram, J. C. (1988). *J. Fish Biol.* **32:** 937-947.

Beckman, B. R, and Zaugg, W. S. (1988). *J. Fish Aquat Sci.* **45:** 1430-1435.

Bergmayer, H.V. (1974). *Academic Press Inc.,* New York, USA, p. 1196.

Bentley, P. J. (1971). *Springer-Verlag,* New York, USA, pp. 220-230.

Berlin, M. and Friberg, L. (1960). *Arch. Environ. Hlth.* **1:** 478-486.

Bhai, I., Idrees, S. and Nath, N. (1977). *Proc. Soc. Exp. Biol. Med.* **138***:* 138-597.

Biesinger, K. E. and Christensen, G. M. (1972). *J. Fish Res. Bd. Can.* **29:** 1691-1700.

Birchall, J. D., Exley, C., Chappell, J. S. and Phillips, M. J. (1989). *Nature* **338:** 146-148.

Birchall, J. D. and Chappell, J. S. (1988). *Clin. Chem.* **34:** 265-267.

Birge, W.J., Hudson, J.E., Black, J.A. and Westennan, A.G. (1980). *'Surface mining and fish wild life needs in the Eastern United States'*

Blasco, J., Puppo, J. and Sarasquete, M. C. (1993). *Mar. Biol.* **115:** 113-118.

Boge, G., Leydet, M. and Houvet, D. (1992). *Oncorhynchus mykiss. Aquat. Toxicol.* **23***:* 247-260.

Booth, C. E., McDonald, D. G., Simons, B. P. and Wood C. M. (1988). *Can. J. Fish Aquat. Sci* **45***:* 1563-1574.

Bouck, G. R. (1972). *Ambloplites repestris. Trans. Am. Fish Soc.* **101:** 488-493.

Bouck, G. R. and Ball, R. C. (1965). *Trans. Am. Fish Soc.* **94:** 363-370.

Britten, J. S. and Blank, M. (1973). *J. Colloid. Interface.* Sci. **43:** 564.

Bruin, A. De. (1976). *Elsevier, North Holland Biomedical Press,* The Netherlands.

Brown, J. A. and Waring, (1996). *Cambridge University Press,* Cambridge, UK.

Burrows, W. D. (1977). *CRC Critical Rev. Environ. Cont.* **7:** 167-216.

Busvine, J. R. (1971). *Cambridge University Press,* Cambridge, UK, pp. 270-276.

Call, D. J., Brooke, L. T., Lindberg, C. A., Markes. T. P., McCauley, D. J. And Poirier. S. H. (1984). Env. Studies, Tech. Rept. 549-238-RT -Wrd.

Cairns, J. (1984). Estimating hazard. *Bioscience.* **30:** *7.*

Caldwell, R. C. (1977). Biochemical effects of pesticide on the Dungeness crab. *EPA-* **600** /3-77 BI: 125.

Cameron, J. N. (1976). *J. Exp. Biol.* **64**: 711-725.

Campbell, R.C. (1988). Statistics for biologists. Cambridge, London, p. 66.

Cengiz S. A. Wissing R. H. Müller.**(2006).** *International Journal of Cosmetic Scoince.* 19.

Chandravathy, V. M and Reddy, S. L. N. (1994). *J. Environ. Biol.* **15(1):** 75-82.

Chavan also Muley, (2014). *Research Gate, Aquaculture Articles.*

Chevalier, G., Gauthier, L. and Moreau, G. (1985). *Can. J. Zool.* **63:** 2062-2070.

Christian, F. A., Jackson, R. N. and Tate, T. M. (1993). *Bull. Environ. Contam. Toxicol.* **51**: 703-709.

Christensen, G. M. (1975). *Appl. Pharmacol.* **32**: 191-197.

Christensen, G. M. (1971) *Chem. Biol. Interactions* **4:** 351-361.

Christensen, G. M., McKim, G. M., Brinos, W. A. and Hunt, E. P. (1972). *Appl. Pharmacol.* **23:** 417-427.

Chitra, T. and Ramana Rao, J.Y. (1986). *Comp. Physiol. Ecol.* **11(2):** 49-53.

Claybrook, D.L. (1983). *Academic Press Inc.,* New York, USA.

Cummins, C. P. (1994). 'Hand book of ecotoxicology'. (Ed.) Calow, P. *Blackwell Scientific Publications,* Oxford, 21-44.

Cyriac, P. J., Antony, A. and Nambisan, P. N. K. (1989). *Bull. Environ. Contam Toxicol.* **43:** 315-320.

Dalela, R. C., Tyagi, A. K., Pal, N. and Verma, S. R (1991). *Water Air Soil Poll.* **15:** 3-9.

Dalwani, R., Dava, J.M. and Datta, K. (1985). *Biochem. Int.* **10***:* 33.

De Silva and Samayawardhena, (2002). *Chemosphere*. **58(9)**:1293-9

Dhavale, D. M., Masurekar, V. B. and Giridhar B. A. (1988). *Bull. Environ. Contam. Toxicol.* **40:** 759-763.

Dhar, A. and Banerjee, P. K. (1983). *J. Vit. Nutr. Res.* **53***:* 349-354.

Driscoll, C. T. and Schecher, W. D. (1988). *Publ. Dekker,* New York and Basel, pp. 59-62.

Driscoll, C. T., Baker, J. P., Bisigni, J. J. and Schofield, C. L. (1980). *Nature Lond* **284: 161-164.**

Drishya MK, Binu Kumari S, Mohan Kumar M, Ambikadevi AP and Aswin B. (2016). *International Journal of Fisheries and Aquatic Studies*. **4(5):** 13-16

Durham, B. W. (1974). *Toxicology of Environmental Pollutants in Industrial Pollution. (Ed.)*

EmiliaStrzyżewska-Worotyńska,JózefSzarek, Izabella Babińska, and Dominika Gulda (2017). *Environ Monit Assess*. **189(12):** 611.

Erickson E. H. and Jones. B. (1939). Links acidity to toxicity for fish.

Evans, D. H. (1987). *Environ. Hlth. Perspect* **71:** 47-58.

Exley, C., Chappell, J. A. and Birchall, J. D. (1991). *J. Theor. Biol.* **151:** 417-428.

Exley, C. and Birchall, J. D. (1992). *Theor. Biol.* **159:** 83-98.

Exley, C., Wicks, A. J., Hubert, R. B. and Birchall, J. D. (1996). *J. Theor. Biol.* **179:** 25-31.

Exley, C., Wicks, A. J., Hubert, R. B. and Birchall, J. D. (1994). *J. Theor. Biol.* **167:** 415-416.

Ferrando, M. D. and Andreu, E. M. (1991). *Bull. Environ. Contam. Toxicol.* **47:** 479-484.

Finney, D. J. (1978). *Griffin Press,* London, UK, p. 508.

Fletcher, C. R (1970). *J. Exp. Biol.* **53(2):** 425-443.

Fletcher, G. L. (1975). *Can. J. Zool.* **53:** 197-206.

Folmar, L. C. (1993). *Environ. Toxicol. Chem.* **12:** 337-375.

Foskett, J. K., Bern, H. A, Machen, T. E. and Conner, M. (1983). *J. Exp. Biol.* **106:** 255-281.

Freda, J, Sanchez, O. A and Bergman, H. L. (1991). *J. Fish Aquat. Sci.* **48:** 2028-2033.

Freeman, R. A. and Everhart, W. H. (1977). *Trans. Am. Fish Soc.* **100:** 644-658.

Galvez, F., Webb, N" Hogstrand, C. and Wood, C. M. (1998). *J. Fish Biol.* **52:** 1089-1104.

Ganrot, P.O. (1986), *Environ. Hlth. Persp* **65:** 363 -441.

Garg, V. K., Garg, S. K. and Tyagi, S. K. (1989). *J. Environ. Biol.* **10(4):** 349-353.

Ghosh, T. K. and Chatterjee, S. K. (1989). *J. Environ. Biol.* **10:** 93-99.

Ghazaly, K. S. (1992). *Water Air Soil Pollute.* **64:** 551-559.

Gill, T. S. and Pant, J. C. (1987). *Water Air Soil Pollute* **35:** 241-250.

Gill, T. S., Tewari, H. and Pande, J. (1992). *Ecotoxicol. Environ. Saf.* **23:** 294-306.

Gill, T. S., Tewari, H. and Pande, J. (1991). *Comp. Biochem. Physiol.* **100:** 501-506.

Gilles, R. and Requeux, A (1983). *Academic Press Inc.,* Newyork, USA.

Goodman, L. S. and Gillman, A.(1970). 'The Pharmacological Basis of Therapeutics'. (Ed.) *Macmillan,* London, UK.

Goss, G. G. and Wood, C M. (1988). *J. Fish Biol.* **32:** 63-76.

Goossernaerts, C, Van Grieken, R., Jacob, W., Witters, H. and Vanderborght, O.(1988). A *Intern. J. Environ. Anal. Chem.* **34***:* 227-237.

Grant, W. C and Root, W. S. (1952). *Physiol. Rev.* **32:** 449.

Greenway, P. (1979). *Academic Press, Inc.,* New York, pp. 117-173.

Gundersen, D. T., Bustaman, S., Seim, W. K. and Curtis, L. R. (1994). *Can. J. Fish Aquat Sci.* **51:** 1345-1355.

Gurr Edward. (1959). *International Information System For The Agricultural Science And Technology*

Handy, R. D. and Eddy, F. B. (1989). *Fish Biol.* **34:** 865-874.

Hall, W. S. and Hall, L. W. (1989). *Bull. Environ. Contam. Toxicol.* **42:** 791-798.

Hagen, E. G. (1959). *Association of Food and Drug officials of the United*

States.

Harman, B. J., Johnson, D. I. and Greenwald, L.(1980). *Trans. Am. Fish Soc.* **109:** 544-551.

Haya, K. and Waiwood B. A., (1983). *J. O. John Wiley and Sons Inc.,* New York, pp. 308-326.

Haux, C. and Larsson, A. (1984). *Aquat. Toxicol.* **5:** 129-142.

Heisler, N. (1982). *Can. J. Zool.* **60:** 1108-1122.

Heisler, N (1986). Amsterdam, The Netherlands, pp.309-356.

Heming, T. A. and Blumhagen, K. A. (1988). *Aquat. Toxicol.* **12:** 125-140.

Hickey, C. R. J. (1976). *N. Y. Fish Game. J.* **23:** 170-175.

Hilmy, A. M., El Domaiaty, N. A., Daabees, A. Y. and Abdel Latife, H. A. (1987). *Comp. Biochem.Physiol.* **87C:** 297-301.

Houston, A. H. and Mearow, K. M. (1982). *Biochem. Physiol. 71A:* 175-180.

Howells, G., Dalziel, T. R. K., Reader, J. P. and Solbe, J. F. (1990). *Chem. Ecol.* **4: 117-173.**

Iwama, G. K., Greer, G. L. and Randall, D. J. (1986). *J. Fish Biol.* ***28:*** 563-572.

Jackim, E., Hamlin, M. and Sonis, S. (1970). *J. Fish Res. Bd. Can.* **27:** 383-390.

Jurss, K. (1980). *Comp. Biochem. physiol.* **65A:** 501.

Karlsson-Norrgren, L., Bjorklund, I., Ljungberg, O., and Runn. P., (1986a). *J. Fish Disc.* ***9:*** 11-25.

Kench, J. E. (1972). *Trans. Roy. Soc. South Africa* **40:** *209.*

King, E.J. and Jagatheesan, K.A. (1959). *J. clin. Path.* **12:** 85.

Kind, P.R.R. and King, E.J. (1954). *J. Clin. Path.* **7:** 322.

Koundinya, P. R., and R. Ramamoorthi, R. (1977). *Experientia* **35:** 1632-1633.

Kumar, A an Barthwal, R.(1991). *Bull. Environ. Contam. Toxicol.***46:** 761-768.

Lamb, D. S. and Bailey, G. C. (1981). *Bull. Environ. Contam Toxicol.* **27:** 59-67.

Larsson, A., Haux, C. and Johansson-Sjobeck, M. (1985). *Ecotoxicol. Environ. Saf.* **9:** 250-280.

Larsson, A., Bengtsson, B. E. and Svanberg. O. (1976). *Cambridge University Press,* Cambridge, U.K., pp. 35-45.

Laurent, P. and Hebibi, N. (1988). *Can. J. Zool.* 3055-3063.

Laurent, J. and McDonald, D.G. (1987). *Can. J. Fish Aquat.* Sci. **44:** 105-111.

Lehninger, A.L. (1982). *Worth Publishers Inc.* New York.

Lehman, 1. and Sturenberg, F. J. (1975). *Gewasser. Abwasser.* **55:** 56-123.

Levinsosn S. S. (1979). Clin. Chem. **22:** 273.

Leland, H. V. (1983). *Environ. Toxicol Chem.* **2:** 353-368.

Lewellen and W.S. King (1964). *Pergamon Press.* London. P.107.

Linhardt, K. and Walter, K. (1965). *Academic Press.* Newyork and London.

Lieberherr, M., Grosse, B., Cournot, W. G., Thil, C. L., and Balsun, S. (1982). *Calcif. Tissue. Int.* **34:** 280-284.

Lorz, H. W. and Mc Pherson, B. P. (1976). *J. Fish Res. Bd. Can.* **33:** 2023-2030.

Lock, R. A. C., Cruijsen, P.J. M. and Van Overbeeke, A. P. (1981). *Comp. Biochem. Physiol.* **69C:** 151-159.

Lowe-Jinde, L. and Nimmi, A.J. (1983). *Can.* J. *Zool.* **61:** 396-402.

Lydersen, E., Salbu, B., Poleo, A. B. S. and Muniz, I. P.(1991). *Water Resour. Res.* **27:** 351-357.

Martin, R. B. (1994). *Acc. Chem. Res.* **27: 204 - 210.**

Mason, B. and Moore, C.B. (1982). *Principles of Geochemistry, Wiley,* New York.

Mayer, F. L., Versteeg, D. J., McKee, M. J., Fohnar, L. C., Graney, R. L., McCume, D. C. and Rattner, B. A. (1992). *Lewis Publishers,* Boca Raton, USA, pp. 5-85.

McCarthy, L. S., and Houstan, A. H. (1976). *J. Fish Biol.* **9:** 11-19.

McDonald, D. G. (1983b). *J. Exp. Biol.* **102:** 123-140.

McDonald, D. G. (1983a). *Can. J. Zool.* **61:** 691-703.

McDonald, D. G. and Wood, C. M. (1981). *J. Exp. Biol.* **93:** 101-118.

McDonald, D. G. and Milligan, C. L. (1988). *Can. J. Fish Aquat. Sci.* **45:** 1606-1613.

McKim, J. M., Christensen, G. M. and Hunt, E. P. (1970). *J. Fish Res. Bd. Can.* **27:** 1883-1889.

McDonald, D.G., Reader, J.P. and Dalziel T.R.K. (1989). J.A.*Cambridge University Press,* Cambridge, UK, pp. 221-241.

McLeay, D. J. and Gorden, M. R. (1977). *J. Fish Res. Bd. Can.* **34:** 2164-2175.

Metelev, V.V. (1983). *Amerind Publishing Co. Pvt. Ltd.,* New Delhi, India, p. 175.

Mount, D. I. and Warner, R. E. (1965). *Publ. Hlth. Serv. Publis. Wash. 999-WP-* **23:** 16.

Müller Ewald and Rita Triebskorn (2001). *Journal of Aquatic Ecosystem Stress and Recovery.* **8** pp 241–260

Muthukumaravel also Rajaraman (2013). *J. Pure Appl. Zool.,* **1(2):** 122-126, 2013

Nagel, R. (1993). *VCH Publishers, Weinheim,* Germany, pp. 147-156.

Narain, A. S. and Srivastava, P. N. (1989). *Bull. Environ. Contam. Toxicol.* **43:** 627-634.

Nath, R. and Banerjee, V. (1995). *J. Freshwater Biol.* **7(4):** 267-268.

Nelson, D.J. and Rani, G.D. (1992). *J. Ecotoxicol. Environ. Monit.* **2(3):** 221 -223.

Nelson, D.A. and Morris, M. W. (1989). *Saunders Company,* Philadelphia, USA, pp. 578-625.

Neff, J. M. (1985). *American Society for Testing Materials. Special Technical Publication. 854:* 155-183.

Nemcsok, J. and Boross, L. (1982). *Acta Biol. Hung.* **33:** 23-27.

Neville, C. M. and Campbell, P. G. C. (1988). *Water Air Soil Pollut.* **42:** 311-327.

Newman, CH Jagoe (1994)- *Bioavailability: Physical, chemical*

Nikinmaa, M. (1982). *Comp. Biochem. Physiol.* **71 A:** 353-356.

Palanichamy, S. and Baskaran, P. (1995). *J. Ecotoxicol. Environ. Monit.* **5(2):** 131-138.

Pamila, D., Subbaiyan, P.S. and Ramaswamy, M. (1990, 1991), *Indian J. Environ. Hlth.* **33(2):** 218-225.

Panchalingam., K, Sachedina., S., Pettegrew, J. W and Glonek, T (1991). *Int. J. Biochem. Vol.* **23:** 1453-1469.

Phillips, M. J. and Saleh, M. A. M. (1988). *J. L.* **1988 No. 15,** pp 489-496.

Pickering, A. D., Pottinger, T. G. and Christie, P. (1982). *J. Fish Biol.* **20:** 229-244.

Pickering, A. D. (1993). *Fish Res.* **17:** 35-50

Plaa, (1982). *Wiley Interscience Publications. John Wiely and Sons,* Newyork.

Playle, R. S., Goss, G. G. and Wood, C. M. (1989). *J. Comp, Physiol.* ***159:*** 539-550.

Playle, R. C. and Wood, C. M. (1990), *Can.* J *Fish Aquat. Sci.* **47:** 1558-1561.

Poleo, A. B. S. and Muniz, I.P. (1993). *Environ. Biol. Fish.* **36:** 193-203.

Poleo, A. B. S. (1995). *Aquat. Toxicol.* **31:** 347-356.

Poleo, A. B. S., Lydersen, E., Rosseland, 8.0., Kroglund, F., Salbu, B., Vogf, R.D. and Kvellestad. A. (1994). *Water Air Soil Pollut.* **75:** 339-351.

Pratap, H. B., Fu, H., Lock, R. A. C. and Wendelaar Bonga, S. E. (1989). *Arch. Environ. Contam. Toxicol.* **18:** 568-575.

Ramade, F. (1987). Chichester, Great Britain, pp. 1-58.

Ramalingam, K. and Ramalingam, K. (1982). *Proc. Indian Acad. Sci. (Anim. Sci.)* **91:** 501-505.

Ramesh, M. (1995). Ph.D. Thesis, *Bharathiar University,* Coirnbatore, India.

Rattner, B. A., Michael, S. D., and Altland, P. D. (1983). *J. Appl. Physiol. Respir. Environ. Exerc. Physiol.* **55(5):** 1408-1412.

Rao, K. R., Kulkarni, D. A., Pillai, K. S. and Mane, U. H. (1987). *Proc. Nat. Symp. Ecotoxic.* 13-20.

Rao AM[1], Jorio A, Pimenta MA, Dantas MS, Saito R, Dresselhaus G, Dresselhaus MS. (2000). *Phys Rev Lett.* **16; 85(16):**3545.

Reid, S. D., Mc Donald, D. O. and Rhem, R. R. (1991). *Can. J. Fish Aquat. Sci.* **48 (10):** 1996-2005.

Rosseland, B.O. and Staurnes, M. (1994). *John Wiley and Sons Ltd.,* pp. 227-246.

Rosseland, B. O., Blakar, I. A., Burger, A., Kroglund, F., Kvellestad, A., Lydersen, E., Oughton, D. H., Salbu, B., Staurnes, M. and Vogt, R. (1992). *Environ. Pollut.* **78:** 3-8.

Ruparelia, S. *G.,* Verma, Y., Saiyed, S. R. and Rawal, U. M. (1990, 1991 and 1992) *Bull. Environ. Contam, Toxicol.* **45:** 305-312.

Rusia, V. and Sood, S. K. (1992). *Tata Mc Graw Hill Publishing Company Limited.* New Delhi, pp. 252-258.

Saad, M. A. H., Ezzat, A. and Shabana, M. B. (1973). *Water Air Soil Pollute.* **2:** 171-179.

Sadler, K. and Lynam, S. (1987). *J. Fish Bio.* **31:** 209-219.

Sakthivel, M. (1988). *Proc. Ind. Acad. Sci. (Anim. Sci.)* **97:** 363-366.

Santhakumar, M., Balaji, M., Ramudu, K. (2001). *J. Environ. Biol.*, **22**, 87-90.

Sastry, K. V. and Subhadra, K. (1985). *Environ. Res.* **36:** 32-45.

Sayer, M. D J., Reader, J. P. and Dalziel, T. R. K. (1993). *Rev. Fish Biol. Fish.* **3:** 95-132.

Schofield, C.L and Trojnar, J.R. (1980). *Plenum Press,* New York, pp. 341-362.

Sen, *G.,* Behera, M. K. and Patel, P. (1992). *J. Ecotoxicol. Environ. Monit.* **2(2):** 89-92.

Sellers, C. M. Jr., Heath, A. G. and Bass, M. L. (1975). *Water Res.* **9(4):** 401-408.

Shiosaka, T., Okuda, H. and Fugi, S. (1971). *Biochem. Biophys. Acta.* **246:** 171-183.

Siegel, N., Cughlin, R. and Haug, A. (1983). *Biochem. Biophys. Res. Commun.* **115:** 512-517.

Singh, H. (1995). *J. Environ. Biol.* **16(4):** 339-341.

Sivakumari, K. (1998). Ph.D. Thesis, *Bharathiar University,* Coimbatore, India.

Soivio, A. and Oikari, A. (1976). *J. Fish Biol* **8:** 397.

Soivio, A. and Nyholm, K. (1973). *Aquaculture.* **2:** 31-35.

Sprague, J. B. (1971). *American Society for Testing Materials,* Philadelphia, USA, pp. 6-28.

Srivastava, J. and Srivastava, A. K. (1984). *The Academy of Environmental Biology,* Muzaffarnagar, India. pp. 37-44.

Staurnes, M., Blix, P. and Reite, O. B. (1993). *Can. J. Fish Aquat. Sci.* **50:** 1816-1827.

Staurnes, M., Sigholt, T. and Reite, O.B. (1984). *Experientia* **40:** 226-227.

Strickland, R.D., Freeman, M.L. and Gurule, F.F. (1961). *Anal. Chem.* **33.**

Svobodova, Z. (1982). *Bull. VURH. Vodnany. 2:* 26-28.

Tewari, H., Gill, T. S. and Pant. J. (1987). *Bull. Environ. Contam. Toxicol.* **38:** 748-752.

Thatheyus, A.J., Selvanayagam, M., & Raja, S. (1992). *Indian Journal of Environmental Health.* **34(3),** 236-238.

Thomsen, A. Korsgaard, B. and Joensen, J. (1988). *Aquat. Toxicol.* **12:** 291-300.

Thurberg, F.F., Calabrese, A, Gould, E., Greeg, R.A, Dawson, MA and Tucker, R.K. (1977). *Academic Press,* New York. p.175.

Tietge, J. E., Johnson, R. D. and Bergman, H. L. (1988). *Can. J. Fish Aquat. Sci.* **45:** 1643-1648.

Tort, L. and Torres, P. (1988). *J. Fish Biol.* **32:** 277.

Torreblanca .J. Del Ramo., JA. Arnau (1989). True Element Research.

Trinder, P. (1969). *Ann. Clin. Biochem.* **6:** 24.

Tucker, R. K. and Matte, A. (1980). *Bull. Environ. Contam. Toxicol.* **24:** 847-852.

Ultsch, G. R. and Gros, G. (1979). *Comp. Biochem. Physiol.* **A 62:** 685-689.

Umminger, B.L. and Gist. D.H. (1970). *Comp. Biochem. Physiol.* **44:** 967-977.

Van der Putte, I., Vander Galien, W. and Strik J. J. T. W. A. (1982). *Ecotoxicol. Environ. Saf.* **4:** 191-194.

Vallee, B. L. and Ulmer, D. D. (1972). *Ann. Rev. Biochem.* **41:** 91-128

Verma, S. R., Jain, M. and Tonk, I. P. (1983). *Toxicol. Lett.* **16:** 305-309.

Verity. M. A and Reith. A. (1967). *Biochem.* J. **105:** 685-690.

Versteeg. D. J. and Giesy, J. P. (1986). *Arch. Environ. Contam. Toxicol.* **14:** 631-640.

Vijayan, M. M. and Moon, T. W. (1992). *Can. J. Fish Aquat. Sci.* **49:** 2260-2266.

Vuorinen, P. J., Vuorinen, M. and Peuranen, S. (1990). *Springer - Verlag,* Berlin, 941-961.

Vuorinen, P. J. and Vuorinen, M. (1991). *Finnish Fish. Res.* **12:** 125-133.

Vuorinen, P. J., Vuorinen, M. and Peuranen, S. (1992). *Environ. Pollut.* **78:** 19-27.

Waring, C.P. and Brown, J. A. (1997). *Gen. Comp. Endocrinol.* **106(1):** 120-126.

Waring, C. P. and Brown, J. A. (1995). *Fish Physiol. Biochem.* **14(1):** 81-91.

Watson, T. A.and Beamish, F. W. H. (1980). *Comp. Biochem. Physiol.* **66C:** 77-82.

Webb, P. W. and Brett, J. R. (1973). *J. Fish Res. Bd. Can.* **30:** 499-507.

Wedemeyer, G. (1972). *J. Fish Res. Bd. Can.* **29:** 1780-1783.

Westfall, B. A. (1945). *Ecol.* **26:** 283-287.

Williams, E. M. and Eddy, F. B (1988). *Aquat Toxicol.* **13(1):** 13-28.

Witters, H. E., Van puymbroeck, S., Stouhart, A. J. H. X. and Wendelaar Bonga, S. E. (1996). *Environ. Toxicol.* **15:** 986-996.

Witters, H. E., Van Puymbroeck, S., Van Den Sande, I. and Vanderborght, O. L. J. (1990). *J. Comp.Physiol. 160B:* 563-571.

Witters, H. E. (1986). *Aquat. Toxicol.* **8:** 197-210.

Wood, C. M. and Caldwell, F. H. (1978). *J. Exp. Zool.* **205:** 301-307.

Wood, C. M., Playle, R. c., Simons, B. A., Goss, G. G. and McDonald, D. G. (1988)a. *Can. J. Fish Aquat. Sci.* **45(9):** 1575-1586.

Wood, C. M., Wheatly, M. G. and Hobe, H. (1984). *Respir. Physiol.* **55:** 175-192.

Yadwad, V.B., Kallapur, V.L. and Basalingappa, S.(1990). *Bull. Environ. Contam. Toxicol.* **44:** 585-589.

Yang, P. and Randall, D.J. (1996). *Bull. Environ. Contam. Toxicol.* **56:** 788-795.

Yilmaz P[1], Kottmann R, Field D, Knight R, Cole JR, Amaral-Zettler L, Gilbert JA, (2011). *Nat Biotechnol.***29**(**5**):415-20.

Youson, J.H. and NeVille, C.M. (1987). *Can .J. Zool.* **65:** 647-656.

Young, D.C. ***et al., (*****1975).** *Clin. Chem* **21:** 12.

Zaug, W. S. (1982). *Can. J. Fish Aquat. Sci.* **39:** 215-217.

the length of by large number of blood cells of the minor gill lamllae. Muthukumaravel also Rajaraman, (2013) establish so as to the poisonous consequence of chromium scheduled the histology of gill also liver of *Labeo rohita* fingerlings going on different publicity.

Yilmaz *et al.* (2011) experimental the histophathological changes into the *Leuciscus cephalus.* They reported to fingerlings showing in the direction of $CdSO_4$ be exposed hydropic deterioration plus necrosis of the minor lamellar epithelia also inflammation of chloride cells. While related results within Cd publicity be seen into additional studies scheduled dissimilar species of fingerlings, various other histopathological changes be in addition report. Publicity of mercuric chloride preserves reason strict pathological alteration into gills also liver of *Cyprinus mrigala.* The main changes into the liver consist of defeat of cellular structural design, necrosis during hepatocytes also addition of fat inside parenchyma cell. The gills show lamellar collapse, epithelial stimulating also necrotic changes into intercellular epithelial cells (Chavan also Muley, 2014).

SUMMARY

- The 24 hours Median lethal concentration of aluminium sulphate on the way to fish is 44.90 ppm in addition to the equal for 96 hours was 26.42 ppm; representative to the metal is poisonous in the direction of fish.

- Experiments are conducted used for a stage of 35 days among weekly sample regularity. No humanity was experimental during the investigational period. Winning completion of the predetermined revelation periods of 7, 14, 21, 28 as well as 35 days, 20 fishes were taken

- Totaling of Median lethal concentration of aluminium sulphate cheap the pH of the analysis water from 7.5 ± 1 to 4.9 ± 1. During this pH variety polymerisation of monomeric aluminium hydroxide variety to upper molecular variety might happen which be very toxic on the way to fish.

- Statement of toxic aluminium polymers going on the gill exterior strength include cause structural alteration such as edema, lamellar synthesis of gills also plentiful discharge of mucus cause severe congestion of the interlamellar seats. The more than alter strength include resulted into impair gaseous exchange within gills most important on the way to mortality of fish for the duration of acute management.

- Na^+ in addition to Cl^- ions be condensed within the plasma of fish showing to acute also sub-lethal concentration of aluminium. This might include be due on the way to the statement of aluminium polymers going on the gills top in the direction of decrease into ionic replace in favor of transfer, payable on the way to the reserve of active ion arrival, stimulus of inactive efflux during paracellular channel or payable to the reserve of transport enzyme inside gills.

- Elevation into plasma potassium within the current study rnay have been due on the way to split of cell membranes cause discharge of cellular potassium within to the plasma decrease in the extra-cellular gap or reduce inside the plasma capacity.

- Blood pH also HCO_3^- level of fish treat by acute aluminium toxicity decrease considerably. Though into sub-lethal management the higher than change be not important. Surplus H^+ ion formed for the duration of aluminium hydrolysis, aluminium induce hypoxia payable in the direction of impair gaseous exchange irritating lactic acid, PCO_2 addition, inhibition of acid-base significant ion exchange strength be the possible causes of blood pH refuse experimental into the current study. Decline of HCO_3^- level into the plasma inside the current study might include be due in the direction of reserve of carbonic anhydrous enzyme.

- Fish showing on the way to together acute also sub-lethal study exhibit cheap levels of haemoglobin, haematocrit also RBC count than so as to the control. This might have been due on the way to trouble inside the fluid volume balance, reserve of erythropoiesis, erythroclasia or interfering of hemometabolism by means of aluminium ions. Leucocytosis experimental into the acute also sub-lethal treatment is almost certainly due toward stimulated immunological method aligned with aluminium strain. Enlargement of red blood cells might have been the reason in support of the major changes in MCHC also MCV values within sub-lethal study.

- The reduce in the plasma protein content during the present study acute in addition on the way to sub-lethal treatment might have been in the direction of rapid utilisation of protein improved proteolysis, decrease in the protein synthesis in the fish below aluminium strain. All through sub-lethal management hypoxia might have been responsible used for the increase of plasma protein satisfied which might act because an osmoeffector subdivision subsequent in the direction of ion trouble caused in aluminium toxicity.

- Plasma glucose be elevated into both acute also sub-lethal treatment. This might include been due in the direction of improved plasma cortisol level into the investigational fish resultant within glycogenolysis or gluconeogenesis on the way to present extra power for the duration of the times of metabolic strain.

- Gill in addition to plasma Na^+, K^+, ATPase action of fish was reserved in a1uminium toxicity both into acute also sub-lethal study. This might include been due in the direction of straight reserve of enzyme in aluminium ions substitute of common co-factor Mg^{2+} ions of the enzyme, compulsory of the metal ion by means of the phosphate grouping of the enzyme or 'Chloride cell' injure. The raise in the plasma Na^+, K^+, ATPase into the sub-lethal action behind 21st days might have been payable to 'hormesis' (Yang plus Randall, 1996) which might be a reimbursement development toward inhibitory challenge created in aluminium.

- Together Acid in addition to alkaline phosphatases actions bigger in the investigational fish which might contain is due in the direction of lysosomal membrane injure cause leakage of cellular enzyme into plasma. But a reduce inside the plasma ALP activity be record into the sub-lethal action. This might be recognized to straight enzyme reserve in aluminium metal which be reported on the way to include a high similarity communication by means of calmodulin, the nonnal activators in the enzyme (Ganrot, 1986).

- The architecture of gill of *Cyprinus carpio var. communis* shows the noticeable descent. Fusions of minor lamellae be and experimental into several region. Alteration curling also clubbing into the tips of the minor lamellae is in addition experimental for the duration of toxicity study behind 28 days.

BIBLIOGRAPHY

Athikesavan, S., Vincent, S., Ambrose, A., & Velmurugan, B. (2006). *Journal of Environmental Biology.* **27: pp** 391-395.

Alabaster, J. S. and Lloyd, R. (1985). *Butterworth Scientific Publications,* London.

Alexander, J. B. and Ingram, G. A. (1980). *J. Fish Biol.* **16:** 115-122.

Apha. (1971). *American Public Health Association,* New York, USA, p. 874.

Apha, Awwa also Wpcf. (1985). *American Public Health Association,* Washington, USA.

Atland, A. (1998). *Environ. Biol. Fish* **53:** 331-345.

CPSIA information can be obtained
at www.ICGtesting.com
Printed in the USA
LVHW080002061222
734629LV00016B/1231